The Engineering Dynamics Course Companion, Part 1: Particles

Kinematics and Kinetics

Complete Supplemental Video Playlist

https://www.youtube.com/playlist?list=PL5aZISlMu3kJnmZ7VX8w3bOLH0SBGg2xn

Synthesis Lectures on Mechanical Engineering

Synthesis Lectures on Mechanical Engineering series publishes 60–150 page publications pertaining to this diverse discipline of mechanical engineering. The series presents Lectures written for an audience of researchers, industry engineers, undergraduate and graduate students.

Additional Synthesis series will be developed covering key areas within mechanical engineering.

The Engineering Dynamics Course Companion, Part 1: Particles: Kinematics and Kinetics
Edward Diehl

ISBN: 978-3-031-79676-0   paperback
ISBN: 978-3-031-79677-7   ebook
ISBN:978-3-031-79678-4   hardcover

DOI 10.1007/978-3-031-79677-7

A Publication in the Springer series
SYNTHESIS LECTURES ON MECHANICAL ENGINEERING

Lecture #26
Series ISSN
Print 2573-3168   Electronic 2573-3176

Newtdog and Wormy are registered trademarks of Edward James Diehl.

The Engineering Dynamics Course Companion, Part 1: Particles

Kinematics and Kinetics

Edward Diehl
University of Hartford

SYNTHESIS LECTURES ON MECHANICAL ENGINEERING #26

ABSTRACT

Engineering Dynamics Course Companion, Part 1: Particles: Kinematics and Kinetics is a supplemental textbook intended to assist students, especially visual learners, in their approach to Sophomore-level Engineering Dynamics. This text covers particle kinematics and kinetics and emphasizes Newtonian Mechanics "Problem Solving Skills" in an accessible and fun format, organized to coincide with the first half of a semester schedule many instructors choose, and supplied with numerous example problems. While this book addresses Particle Dynamics, a separate book (Part 2) is available that covers Rigid Body Dynamics.

KEYWORDS

dynamics, particle kinematics, particle kinetics, Newtonian mechanics

Contents

Acknowledgments

This course companion is the result of a decade of teaching Dynamics in close cooperation with several brilliant and dedicated engineering educators. I would like to sincerely thank my colleagues and mentors for their assistance and inspiration. Here I acknowledge their contribution to this effort and my career as an educator in reverse chronological order.

I'm grateful to my fellow faculty at the University of Hartford, many of whom reviewed the manuscript and offered extremely valuable and insightful feedback and corrections. These include Dr. Cy Yavuzturk, Dr. Mark Orelup, Dr. Mary Arico, Dr. Taka Asaki, Professor Phil Faraci, and Dr. Chris Jasinski. I'm indebted to my Ph.D. advisor, friend, and mentor, Dr. Jiong Tang, for his support and encouragement to publish a work that mattered to me. I'm forever thankful to my friends and colleagues at the United States Coast Guard Academy for supporting me during and providing the opportunity to transition from a practicing engineer to an educator. These include Dr. Todd Taylor, Captain Mike Corl, Dr. Elisha Garcia, Lieutenant (Ret) Sean Munnis, Commander Nick Parker, Dr. Tom DeNucci, Dr. Susan Swithenbank, Commander John Goshorn, and Lieutenant Commander J.J. Schock. The close working relationship of these instructors in which we shared notes, examples, and exam problems heavily influenced the content of this book, and many of the problems within are adaptations of this group effort. I'd like to acknowledge Sean Munnis in particular as the person who dubbed Sir Isaac Newton "Newtdog" and encouraged me more than anyone to draw him as a cartoon character and develop this into a book. Before I joined academia, I was a working engineer and I'm grateful for my former colleagues at Seaworthy Systems and General Dynamics, but especially my mentor, the late Bill McCarthy, who pushed me and inspired me to be a better engineer and better writer. Special thanks to the late Professor Don Paquette of the United States Merchant Marine Academy, my Statics, Dynamics, and Machine Design professor. He inspired me to become a teacher, and I've endeavored to follow in his footsteps.

And in real life, I'm so very thankful to my incredibly supportive wife, Lori Dappert Diehl, who actually read this book. Thank you, Lori, for always making me laugh and never letting me give up. Lastly, I'd like to acknowledge my older brother, the late James Harold Diehl, whose communication limitations have obliged me to communicate and whose resilience inspires me to persevere through the relatively minor inconveniences of life.

Edward Diehl
August 2020

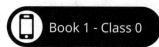

Book 1 - Class 0

https://www.youtube.com/watch?v=WvowYa_3OQg

CLASS 0

Introduction

B.L.U.F. (Bottom Line Up Front)

- Dynamics is the study of motion.

- Kinematics and Kinetics:

    - Kinematics: the description of motion, ignoring the cause of the motion.

    - Kinetics: the interaction of loading and motion on objects with mass.

- Categories of objects:

    - Particles: objects treated as point masses since their size and shape isn't important.

    - Rigid Bodies: objects whose size and shape m their rotation is important to how they move.

- Dynamics courses are often broken down into four parts:

    - Particle Kinematics, Particle Kinetics, Rigid Body Kinematics, Rigid Body Kinetics.

0.1   ABOUT THE BOOK

This is the first part of a two-part "course companion" to assist undergraduate engineering students taking a first course in Dynamics. Part 1 deals with the dynamics of particles, while Part 2 covers the dynamics of rigid bodies. Your course companions are "Newtdog and Wormy" (Figure 1) who will guide you through Newtonian Dynamics with plenty of examples and images especially geared towards visual learners.

Many engineering majors typically take Dynamics in their Sophomore year after completing Statics, and this is one of the most challenging transitions, requiring considerable personal growth in the way problems are approached and processed. This book is meant to help with that transition and serve as a "course companion," a complementary resource a struggling student can refer to when frustrated.

Figure 1: Portrait of Sir Isaac Newton by Godfrey Kneller with your course companions: Newt-dog and Wormy (©E. Diehl).

Why is Dynamics so difficult? Many of the problem types require students to think differently than they're used to: rely less on step-by-step procedures and instead recognize the nature of a problem and navigate to a solution using concepts. Sometimes problems require working backward or applying logic to generate an "ah-ha" moment, when the lightbulb goes off and the path to a solution becomes clear. Students often describe some Dynamics assignments as "trick problems." This is true in a way: the solution will seem obvious once revealed. A good problem solver doesn't need to have worked through an identical problem in order to solve a new problem they've never seen. Instead, with problem solving experience, they develop a skill to pick it apart, identify the underlying principles, and formulate a path forward. Sometimes this is like a maze, where going down one path leads to a dead end. Problem solvers know to reverse course a bit, reevaluate, and try a new approach. This text is intended to be your companion on that journey to developing "ah-ha" skills.

Because this is a "course companion," the book is written in a relatively casual tone compared to most textbooks (note the frequency of the pronoun "we") and includes Sir Isaac Newton as a cartoon to add some levity to this often-dreaded course. The cartoons are intended to also serve as "visual mnemonics." That is, they are meant to be memorable with an aspect of them associated with particular concepts as they're presented. Solving problems in dynamics requires recognizing the nature of a problem, identifying the key concepts, and applying a solution strategy. The middle part is where these cartoons can help, especially if one can think "oh, this is just like when _____." The blank being an aspect of the cartoon.

The 2 parts of this course companion consist of 12 "classes," each coinciding with the typical 2-class-a-week schedule of a semester-long Dynamics course. A common complaint of Dynamics students is not having enough examples or that the available examples are much easier than the homework. Therefore, the examples within each class are progressively longer and more challenging. Some textbooks skip steps within the example solutions, so this course companion attempts to work through the solutions in exhaustive detail. Example exam questions are also included in the appendices to provide opportunity for additional problem-solving practice. This course companion is also intended to assist instructors seeking inspiration for their own examples, homework problems and exam problems.

0.2   NEWTDOG AND WORMY: YOUR COURSE COMPANIONS

"Newtdog" is a silly nickname for Sir Isaac Newton intended to make him less intimidating. Based on quotes, his own writings, and biographies, Sir Isaac Newton seems to have been a down-to-Earth regular guy who was inquisitive and humble. He said: "If I have seen further than others, it is by standing upon the shoulders of giants." and "To myself I am only a child playing on the beach, while vast oceans of truth lie undiscovered before me."

Newtdog is drawn to seem friendly, adventurous (just as Sir Isaac Newton was revolutionary), and a little bit of a dandy with his powered wig, frilly cuffs, long coat, and buckled shoes. Newtdog's buddy is "Wormy" who lives in the iconic apple that apocryphally led Newton to "discover gravity." Wormy is often just along for the ride and a little nervous about Newtdog's enthusiasm and adventurous spirit.

0.3   BOTTOM LINE UP FRONT (B.L.U.F.)

Every chapter begins with a "Bottom Line Up Front" (B.L.U.F.) consisting of bulleted items of the contents with very brief summaries and/or equations. The purpose is to introduce you to the essentials of the topic(s) covered and serve as a quick reference for later use when flipping through the book to search for content. In a classroom environment the BLUF provided at the beginning of class helps get the students prepared for what they're about to learn. Students using this course companion should read the BLUF just before class (at a minimum) so you're on the lookout for this information. The BLUF only takes a few seconds, so it's easy.

0.4   KINEMATICS VERSUS KINETICS

Dynamics can be organized into two parts: Kinematics and Kinetics. It's useful to memorize the definition of these terms to help organize the approaches we'll take.

Kinematics is the description of motion without regards to why it's happening and studies the relationships among time, position, velocity, and acceleration of an object. Kinematics is also often described as "the geometry of motion." We'll begin with particle kinematics since it is perhaps the simplest Dynamics broad topic and introduces many fundamental sub-topics which are useful to build upon.

Kinetics investigates why motion occurs and the interaction between loads and mass. The approach taken here falls into the category of Newtonian Mechanics since it's based on the principals described by Sir Isaac Newton in 1687. This is considered Classical Mechanics which also includes Lagrangian (1788) and Hamiltonian (1833) Mechanics that are reformulations of Newton's approach. You've likely also heard of Quantum Mechanics which shows that Classical Mechanics breaks down on the atomic and sub-atomic level. There are many other methods of studying motion, but Newtonian remains the cornerstone of an engineering education.

Another categorization of Dynamics topics is by Particles and Rigid Bodies. Particles, covered in Part 1, are point masses whose shapes, aren't considered important enough to be included in the analysis. Rigid Bodies, covered in Part 2, have a shape significant enough to include the effect of rotation into the analysis, but their flexibility isn't sufficient to influence the results.

0.5   COURSE BREAKDOWN

The course companion (Parts 1 and 2) are organized to follow the classes of a typical Dynamics course. Part 1 covers Particle Kinematics and Kinetics, and Part 2 covers Rigid Body Kinematics and Kinetics.

- Part 1:
  - Kinematics of Particles
    1. Rectilinear Motion of Particles
    2. Special Cases and Relative Motion
    3. Curvilinear Motion of Particles and Projectile Motion (Rectangular)
    4. Non-Rectangular Coordinates (Path)
    5. Non-Rectangular Coordinates (Polar)
  - Kinetics of Particles
    6. Newton's Second Law in Rectangular Coordinates
    7. Newton's Second Law in Path and Polar Coordinates
    8. Work and Energy, Conservation of Energy
    9. Work and Energy, Conservation of Energy (Part 2)
    10. Impulse and Momentum
    11. Direct Central Impact

12. Oblique Central Impact

- Part 2:
  - Kinematics of Rigid Bodies
    13. Translation and Fixed Axis Rotation
    14. General Plane Motion, Absolute and Relative Velocity
    15. Instantaneous Center of Rotation
    16. General Plane Motion: Acceleration
    17. General Plane Motion: Acceleration (Part 2)
    18. Analyzing Motion w.r.t. a Rotating Frame (Coriolis)
  - Kinetics of Rigid Bodies
    19. Mass Moment of Inertia
    20. Newton's Second Law in Constrained Plane Motion
    21. Newton's Second Law in Translation and Rotation Plane Motion
    22. Energy Methods
    23. Momentum Methods
    24. Eccentric Impact

The topics are broken down in this manner to coincide with a 23 per week, 14-week semester arrangement. Given these 28 possible class periods and subtracting 3 periods for exams the last day of class for review, 24 classes are available to introduce topics. Note there are two classes (9 and 17) that repeat the previous class topic. These are included to give more emphasis to those topics that might otherwise be too much information to absorb in one class. Experience has shown this to be a practical schedule for this level of an Engineering Dynamics course.

Table 0.1 presents a suggested course schedule.

0.6   EQUATION SHEET

Table 0.2 presents a suggested equation sheet for instructors who choose to provide one during exams rather than have students make their own. It is purposefully limited to only two sheets of equations and does not include every permutation of the equations but enough to avoid students' having to memorize formulae. Students who have the option to write their own equation sheet should refer to this to ensure they've covered all the essentials.

0.7   TEXTBOOKS AND REFERENCES

As this is a course companion, it is likely your instructor will assign another textbook. The following is a short list of especially well-written textbooks that are often adopted by instructors.

Table 0.1: Course schedule

Week	Class	Topic
1	1	Kinematics of Particles - Rectilinear Motion of Particles
	2	Kinematics of Particles - Special Cases: Relative and Dependent Motion
2	3	Kinematics of Particles - Curvilinear Motion of Particles (Rectangular)
	4	Kinematics of Particles - Non-Rectangular Components (Path)
3	5	Kinematics of Particles - Non-Rectangular Components (Path)
		Exam 1 (Covering Classes 1–5)
4	6	Kinetics of Particles - Newton's Second Law in Rectangular Coordinates
	7	Kinetics of Particles - Newton's Second Law in Path and Polar Coordinates
5	8	Kinetics of Particles - Work and Energy and the Conservation of Energy (Part 1)
	9	Kinetics of Particles - Work and Energy and the Conservation of Energy (Part 2)
6	10	Kinetics of Particles - Impulse-Momentum Method
	11	Kinetics of Particles - Direct Impact of Particles and the Conservation of Linear Momentum
7	12	Kinetics of Particles - Oblique Impact of Particles
		Exam 2 (Covering Classes 8–12)
8	13*	Kinematics of Rigid Bodies - Angular Kinematics of Rigid Body Motion
	14	Kinematics of Rigid Bodies - Absolute and Relative Velocity
9	15	Kinematics of Rigid Bodies - Velocity Analysis Using the Instantaneous Center of Rotation
	16	Kinematics of Rigid Bodies - Acceleration Analysis (Part 1)
10	17	Kinematics of Rigid Bodies - Acceleration Analysis (Part 2)
	18	Kinematics of Rigid Bodies - Coriolis Acceleration Analysis
11		**Exam 3 (Covering Classes 13–18)**
	19	Kinetics of Rigid Bodies - Mass Moment of Inertia
	20	Kinetics of Rigid Bodies - Newton's Second Law in Constrained Plane Motion
	21	Kinetics of Rigid Bodies - Newton's Second Law in Translating and Rotating Plane Motion
13	22	Kinetics of Rigid Bodies - Rigid Body Work-Energy Method
	23	Kinetics of Rigid Bodies - Rigid Body Impulse-Momentum Method
14	24	Kinetics of Rigid Bodies - Impact of Rigid Bodies
	25	Course Summary and Review for Final Exam
		Final Exam (Covering Entire Semester but emphasizing Classes 19–24)
* Classes 13–24 are covered in *Engineering Dynamics Course Companion, Part 2: Rigid Bodies*		

Table 0.2: Dynamics exam equation sheet (*Continues.*)

Particle Kinematics	Rigid Body Kinematics

Particle Kinematics

Velocity and acceleration in rectilinear motion:

$$\vec{\mathbf{v}} = \frac{d\vec{\mathbf{r}}}{dt} \qquad \vec{\mathbf{a}} = \frac{d\vec{\mathbf{v}}}{dt} = \frac{d^2\vec{\mathbf{r}}}{dt^2} = \vec{\mathbf{v}}\frac{d\vec{\mathbf{v}}}{d\vec{\mathbf{r}}}$$

Uniform translational motion:

$$\vec{\mathbf{r}} = \vec{\mathbf{r}}_0 + \vec{\mathbf{v}}_c t \qquad x = x_0 + v_{x,c} t$$

Uniformly accelerated translational motion:

$$\vec{\mathbf{v}} = \vec{\mathbf{v}}_0 + \vec{\mathbf{a}}_c t \qquad v_x = (v_0)_x + a_{x,c} t$$

$$\vec{\mathbf{r}} = \vec{\mathbf{r}}_0 + \vec{\mathbf{v}}_0 t + \tfrac{1}{2}\vec{\mathbf{a}}_c t^2 \qquad x = x_0 + (v_0)_x t + \tfrac{1}{2} a_{x,c} t^2$$

$$\vec{\mathbf{v}}^2 = \vec{\mathbf{v}}_0^2 + 2\vec{\mathbf{a}}_c(\vec{\mathbf{r}} - \vec{\mathbf{r}}_0) \qquad v_x^2 = (v_0)_x^2 + a_{x,c}(x - x_0)$$

Relative motion of two particles (or points):

$$\vec{\mathbf{r}}_B = \vec{\mathbf{r}}_A + \vec{\mathbf{r}}_{B/A} \qquad x_B = x_A + x_{B/A}$$

$$\vec{\mathbf{v}}_B = \vec{\mathbf{v}}_A + \vec{\mathbf{v}}_{B/A} \qquad v_B = v_A + v_{B/A}$$

$$\vec{\mathbf{a}}_B = \vec{\mathbf{a}}_A + \vec{\mathbf{a}}_{B/A} \qquad a_B = a_A + a_{B/A}$$

Path (tangential and normal) components:

$$\vec{\mathbf{v}} = v_t \hat{\mathbf{e}}_t = (v)\hat{\mathbf{e}}_t$$

$$\vec{\mathbf{a}} = a_t \hat{\mathbf{e}}_t + a_n \hat{\mathbf{e}}_n = \frac{dv}{dt}\hat{\mathbf{e}}_t + \frac{v^2}{\rho}\hat{\mathbf{e}}_n$$

Polar (radial and transverse) components:

$$\vec{\mathbf{v}} = v_r\hat{\mathbf{e}}_r + v_\theta\hat{\mathbf{e}}_\theta = (\dot{r})\hat{\mathbf{e}}_r + (r\dot{\theta})\hat{\mathbf{e}}_\theta$$

$$\vec{\mathbf{a}} = a_r\hat{\mathbf{e}}_r + a_\theta\hat{\mathbf{e}}_\theta = (\ddot{r} - r\dot{\theta}^2)\hat{\mathbf{e}}_r + (r\ddot{\theta} + 2\dot{r}\dot{\theta})\hat{\mathbf{e}}_\theta$$

Rigid Body Kinematics

Rotation about a fixed axis:

$$\vec{\mathbf{v}} = \frac{d\vec{\mathbf{r}}}{dt} = \vec{\omega}\times\vec{\mathbf{r}}, \qquad \vec{\omega} = \omega\hat{\mathbf{k}} = \dot{\theta}\hat{\mathbf{k}}$$

$$\vec{\mathbf{a}} = \underbrace{\vec{\alpha}\times\vec{\mathbf{r}}}_{\vec{\mathbf{a}}_t} + \underbrace{\vec{\omega}\times(\vec{\omega}\times\vec{\mathbf{r}})}_{\vec{\mathbf{a}}_n}, \qquad \vec{\alpha} = \alpha\hat{\mathbf{k}} = \ddot{\theta}\hat{\mathbf{k}}$$

$$a_t = r\alpha, \quad a_n = r\omega^2 \text{ (in one plane)}$$

Angular velocity and angular acceleration:

$$\vec{\omega} = \frac{d\vec{\theta}}{dt} \qquad \vec{\alpha} = \frac{d\vec{\omega}}{dt} = \frac{d^2\vec{\theta}}{dt^2} = \vec{\omega}\frac{d\vec{\omega}}{d\vec{\theta}}$$

Uniform rotational motion:

$$\theta = \theta_0 + \omega_c t$$

Uniformly accelerated rotational motion:

$$\omega = \omega_0 + \alpha_c t$$

$$\theta = \theta_0 + \omega_0 t + \tfrac{1}{2}\alpha_c t^2$$

$$\omega^2 = \omega_0^2 + 2\alpha_c(\theta - \theta_0)$$

Velocity in plane motion:

$$\vec{\mathbf{v}}_B = \vec{\mathbf{v}}_A + \vec{\mathbf{v}}_{B/A} = \vec{\mathbf{v}}_A + \omega\hat{\mathbf{k}}\times\vec{\mathbf{r}}_{B/A}$$

Acceleration in plane motion:

$$\vec{\mathbf{a}}_B = \vec{\mathbf{a}}_A + \vec{\mathbf{a}}_{B/A} = \vec{\mathbf{a}}_A + (\vec{\mathbf{a}}_{B/A})_t + (\vec{\mathbf{a}}_{B/A})_n$$

$$= \vec{\mathbf{a}}_A + \alpha\hat{\mathbf{k}}\times\vec{\mathbf{r}}_{B/A} - \omega^2\vec{\mathbf{r}}_{B/A}$$

Relative Motion <u>on</u> a Rigid Body:

Velocity of a point on a rigid body in plane motion:

$$\vec{\mathbf{v}}_B = \vec{\mathbf{v}}_A + \omega\hat{\mathbf{k}}\times\vec{\mathbf{r}}_{B/A} + \vec{\mathbf{v}}_{rel}$$

Acceleration of a point on a rigid body in plane motion:

$$\vec{\mathbf{a}}_B = \vec{\mathbf{a}}_A + \alpha\hat{\mathbf{k}}\times\vec{\mathbf{r}}_{B/A} - \omega^2\vec{\mathbf{r}}_{B/A} + \vec{\mathbf{a}}_{rel} + 2\vec{\omega}\times\vec{\mathbf{v}}_{rel}$$

Table 0.2: (*Continued.*) Dynamics exam equation sheet

Particle Kinetics	**Rigid Body Kinetics**
Linear momentum of a particle: $\vec{\mathbf{L}} = m\vec{\mathbf{v}}$	Mass center: $m\vec{\mathbf{r}} = \sum_{i=1}^{n} m_i\,\vec{\mathbf{r}}_i$
Angular momentum of a particle: $\vec{\mathbf{H}} = \vec{\mathbf{r}} \times m\vec{\mathbf{v}}$	Moments of inertia of masses and radius of gyration:
Newton's second law: $\Sigma\vec{\mathbf{F}} = m\vec{\mathbf{a}} = \dot{\vec{\mathbf{L}}}$	$$I = \int r^2\,dm \qquad k = \sqrt{\frac{I}{m}}$$
Equations of motion for a particle:	Parallel-axis theorem: $I = \bar{I} + md^2$
Cartesian: $\Sigma F_x = ma_x \qquad \Sigma F_y = ma_y \qquad \Sigma F_z = ma_z$	Equations for the plane motion of a rigid body:
Path coord.: $\Sigma F_t = m\dfrac{dv}{dt} \qquad \Sigma F_n = m\dfrac{v^2}{\rho}$	$$\Sigma F_x = m\bar{a}_x \qquad \Sigma F_y = m\bar{a}_y$$
Polar coord: $\Sigma F_r = m(\ddot{r} - r\dot{\theta}^2) \qquad \Sigma F_\theta = m(r\ddot{\theta} + 2\dot{r}\dot{\theta})$	$$\Sigma M_O = \Sigma(M_O)_{eff} = \bar{I}\alpha + ma_G\,r_{G/O}$$
Principle of work and energy:	Work of a couple of moment M:
$$KE_1 + PE_1 + U_{1\to2} = KE_2 + PE_2$$	$$U_{1\to2} = \int_{\theta_1}^{\theta_2} \vec{\mathbf{M}} \cdot d\vec{\theta}$$
Work of a force: $U_{1\to2} = \int_{A_1}^{A_2} \vec{\mathbf{F}} \cdot d\vec{\mathbf{r}}$	Kinetic energy in plane motion:
Kinetic energy of a particle: $KE = \frac{1}{2}\,mv^2$	$$KE = \tfrac{1}{2}\,m\bar{v}^2 + \tfrac{1}{2}\,\bar{I}\omega^2 = \tfrac{1}{2}\,I_O\,\omega^2$$
Potential energy: $PE_g = mgy,\ PE_{sp} = \frac{1}{2}\,kx^2$	Principle of impulse and momentum for a rigid body:
Power and mechanical efficiency:	**Syst Momenta$_1$ + Syst Ext Imp$_{1\to2}$ = Syst Momenta$_2$**
Power $= \dfrac{dW}{dt} = \dot{U} = \vec{\mathbf{F}} \cdot \vec{\mathbf{v}}$ or $\dfrac{Md\theta}{dt} = M\omega$	Angular momentum in plane motion about mass center:
$\eta = \dfrac{\text{energy (or power) output}}{\text{energy (or power) input}}$	$$\vec{\mathbf{H}}_G = \bar{I}\vec{\omega}$$
Principle of impulse and momentum for particles:	Principle of impulse and momentum for particles:
$$\vec{\mathbf{L}}_1 + \overrightarrow{\mathbf{IMP}}_{1\to2} = \vec{\mathbf{L}}_2$$	$$(\vec{\mathbf{H}}_O)_1 + (\overrightarrow{\mathbf{Ang\ IMP}}_O)_{1\to2} = (\vec{\mathbf{H}}_O)_2$$
$$\Sigma m\vec{\mathbf{v}}_1 + \int_{t_1}^{t_2}\vec{\mathbf{F}}\,dt = \Sigma m\vec{\mathbf{v}}_2$$	$$\Sigma \bar{I}\vec{\omega}_1 + \Sigma m\vec{\mathbf{v}}_1 \times \vec{\mathbf{r}}_{G/O} + \int_{t_1}^{t_2}\overrightarrow{\mathbf{M}}_O\,dt = \Sigma \bar{I}\vec{\omega}_2 + \Sigma m\vec{\mathbf{v}}_2 \times \vec{\mathbf{r}}_{G/O}$$
Oblique central impact:	
$$(v_A)_t = (v_A')_t, \qquad (v_B)_t = (v_B')_t$$	
$$m_A(v_A)_n + m_B(v_B)_n = m_A(v_A')_n + m_B(v_B')_n$$	
$$(v_B')_n - (v_A')_n = e[(v_A)_n - (v_B)_n]$$	

While there are numerous original example problems within this course companion, many of the examples were inspired by problems written by others, especially from these four textbooks. Tables 0.3 and 0.4 are provided to both give appropriate attribution to the original inspiration and to direct instructors to similar problems for homework and exams. Students and instructors are encouraged to explore these problems for alternate arrangements, objectives, and solution approaches. Problems marked with an asterisk indicate that the example was inspired by it and/or is similar enough to be considered an adaptation. Below are the four textbook references.

1 Beer, F. P., Johnston, E. R., Cornwell, P. J., and Self, B. P. 2018. *Vector Mechanics for Engineers: Dynamics*, 12th ed., McGraw-Hill Education, New York.

2 Hibbeler, R. C. 2010. *Engineering Mechanics: Dynamics*, 12th ed., Prentice Hall, Upper Saddle River, NJ.

3 Bedford, A. and Fowler, W. L. 2008. *Engineering Mechanics: Dynamics*, 5th ed., Pearson Prentice Hall, Upper Saddle River, NJ.

4 Tongue, B. H. 2010. *Dynamics: Analysis and Design of Systems in Motion*, 2nd ed., John Wiley & Sons, Hoboken, NJ.

Other references:
Below are additional references used while writing Part 1.

- Newton, I. *Philosophiæ Naturalis Principia Mathematica*, 1687. DOI: 10.5479/sil.52126.39088015628399.

- Meriam, J. L. and Kraige, L. G. 2012. *Engineering Mechanics: Dynamics*, vol. 2, John Wiley & Sons.

- To, Cho W. S. 2018. *Engineering Dynamics. Synthesis Lectures on Mechanical Engineering 2.5*, pp. 1–189. DOI: 10.2200/S00853ED1V01Y201805MEC015

- Nelson, E., Best, C. L., Best, C., McLean, W. G., McLean, W. G., and McLean, W. 1998. *Schaum's Outline of Engineering Mechanics*, McGraw Hill Professional.

- Farrow, W. C. and Weber, R. 1993. *Study Guide to Accompany Engineering Mechanics Dynamics*, Chichester, John Wiley.

- National Council of Examiners for Engineering, 2011. *Fundamentals of Engineering: Supplied-reference Handbook*. Kaplan AEC Engineering.

- Diehl, E. J. Using Cartoons to Enhance Engineering Course Concepts. 2018 ASEE Annual Conference & Exposition. https://peer.asee.org/authors/39810

Table 0.3: Example Problem Reference Correlation for Classes 1–8 (* indicates problem was inspired by)

Class 1					Class 2				
Ex.	[1]	[2]	[3]	[4]	Ex.	[1]	[2]	[3]	[4]
1.1	11.3	12-39	13.10	2.1.13	2.1	11.36	12-29	13.18	2.1.14
1.2	11.13	12-26	13.33	2.1.19	2.2	11.40	12-32	13-82	2.1.22
1.3	11.17	12-18	13.52	2.1.5	2.3	11.52	12-198	-	2.5.5
1.4	11.22	12-16	13.49	2.1.17	2.4	11.55	12-205	-	2.5.8
					2.5	11.59	12-200	-	2.5.25
					2.6	11.56	12-208	-	2.5.28
Class 3					Class 4				
Ex.	[1]	[2]	[3]	[4]	Ex.	[1]	[2]	[3]	[4]
3.1	11.90	12-72	13.87	2.2.2	4.1	-	12-137	13.159	2.2.7*
3.2	11.100	12-104	13.76	2.2.14	4.2	11.143	12-160	13.127	2.4.7
3.3	11.117*	12-218	13.176	2.5.11	4.3	11.15	12-136	13.124	2.4.26
3.4	11.111	12-234	13.162	2.2.16	4.4	11.151	12-121	13.125	2.4.18
Class 5					Class 6				
Ex.	[1]	[2]	[3]	[4]	Ex.	[1]	[2]	[3]	[4]
5.1	11.161	12-168	13.14	2.3.18	6.1	12.15	13-3	14.36	3.1.17*
5.2	11.169	12-187	13.156	2.3.19	6.2	12.14	13-28	14.3	3.1.42
5.3	11.32	12.185	13.157*	-	6.3	12.15*	13-11	14.28	3.1.26
5.4	11.166	12.182*	13.154	2.5.19	6.4	12.22*	13-35	14.4	3.1.45
Class 7					Class 8				
Ex.	[1]	[2]	[3]	[4]	Ex.	[1]	[2]	[3]	[4]
7.1	12.47	13-62	14.133	3.3.9	8.1	13.12	14-4	15.4	4.1.4
7.2	12.27	13-60	14.85	3.3.23	8.2	13.20	14-8	15.26	-
7.3	12.68	13-90	14.97	3.2.5	8.3	13.41	14-66	15.37	4.2.44
7.4	12.71*	-	14.1	3.2.7	8.4	13.12	14-21	15.27	4.1.20*
7.5	12.52	13-71	14.9	3.3.17	8.5	13.58*	14-84	15.127	4.2.13
7.6	12.37	13-83	14.76						

Table 0.4: Example Problem Reference Correlation for Classes 9–12 and Appendix A (* indicates problem was inspired by)

Class 9					Class 10				
Ex.	[1]	[2]	[3]	[4]	Ex.	[1]	[2]	[3]	[4]
9.1	13.69*	14-36	15.52	4.2.25	10.1	13.122	15-10	16.2	3.4.11
9.2	13.45	14-16	15.85	4.2.19	10.2	13.133	15-28	16.20	
9.3	13.51	14-54	-	4.3.36	10.3	13.136	15-22	16.13	3.4.20
9.4	13.54	14-51	15.146	4.3.11	10.4	13.139	15-29	16.39	3.4.23
					10.5	12.91	15-108	16.93	3.5.1
Class 11					Class 12				
Ex.	[1]	[2]	[3]	[4]	Ex.	[1]	[2]	[3]	[4]
11.1	13.146*	15-38	16.55	3.4.17	12.1	13.164*	15-79	16.81	3.8.3
11.2	13.155	15-60	16.78	3.7.6	12.2	13.171	15-77	16.82	3.8.16
11.3	13.157	15-76	16.70	3.7.10	12.3	13.166	15-89	16.79	3.8.17
11.4	13.152	15-39	16.53	3.7.3	12.4	13.186*	15-81	16.136	3.8.22
Appendix A.1					Appendix A.2				
Prob.	[1]	[2]	[3]	[4]	Prob.	[1]	[2]	[3]	[4]
A.1.1	11.98	12-108	13.74	2.2.13	A.2.1	12.14	13-4	14.5*	3.1.36*
A.1.2	11.104	12-94	13.76	2.2.11	A.2.2	12.28	13-28	14.127	3.1.42
A.1.3	11.169*	12-136	13.115	2.3.34	A.2.3	12.36	13-65	14.77	3.3.23
A.1.4	11.166	12-175	13.183	2.3.20	A.2.4	12.37	13-71	14.76*	3.3.6
A.1.5	11.170	12-99	13.144	2.3.31	A.2.5	12.32	13-17	14.130	3.1.56
A.1.6	11.140	12-120	13.117*	2.4.24	A.2.6	13.20	14-14	15.57	4.1.20
A.1.7	11.167*	12-188*	13.141	2.3.21	A.2.7	13.69	14-36	15.59	4.2.25
A.1.8	11.186	12-205	-	2.5.28	A.2.8	13.122	15-10	16.2	3.4.11*
					A.2.9	13.126	15-30	16.12	3.4.43
					A.2.10	13.162*	15-64	16.61	3.7.4
					A.2.11	13.163*	15-59	16.59	3.7.12
					A.2.12	13.166	15-89*	16.83	3.8.19

Book 1 - Class 1

https://www.youtube.com/watch?v=LfxvqVnaD08

C L A S S   1

Rectilinear Motion of Particles

B.L.U.F. (Bottom Line Up Front)

- Fundamental kinematic relationships describe the interrelationships between time, position, velocity, and acceleration.

$$\vec{\mathbf{v}} = \frac{d\,\vec{\mathbf{r}}}{dt} \qquad \vec{\mathbf{a}} = \frac{d\,\vec{\mathbf{v}}}{dt} = \frac{d^2\,\vec{\mathbf{r}}}{dt^2} = \vec{\mathbf{v}}\frac{d\,\vec{\mathbf{v}}}{d\,\vec{\mathbf{r}}}. \tag{1.1}$$

- Determining the motion of a particle moving along a straight line (rectilinear), the vector notation goes away, and is just a positive or negative scalar ("r" can be replaced with any letter, often s, or x, or y, or z)

$$v = \frac{dx}{dt} = \dot{x} \qquad a = \frac{dv}{dt} = \frac{d^2x}{dt^2} = \ddot{x} = v\frac{dv}{dx}.$$

1.1   FUNDAMENTAL KINEMATIC EQUATIONS

The fundamental kinematics equations relate time, position, velocity and acceleration. We'll use them in this chapter in one-dimension (1-D) and later treat them as vectors in two and three-dimensions (2-D and 3-D). We usually think of 1-D as being a straight line, but sometimes we might want to know about motion along a path. Describing the motion of the apple rolling after it leaves Newtdog's hand in Figure 1.1 is an example of 1-D kinematics.

Position is rather obvious but worth a somewhat formal discussion. We need to establish a reference and the distance of an object from it in order to establish its position. For 1-D particles we draw a straight and measure along it to report the position. For 2-D and 3-D we'll use a position vector along with a reference point (origin) in a coordinate system so we'll know exactly where the object is located. We report the position with a coordinate (in 1-D this is often simply x, although s or r are often used). When there is a change in position, we call that a "displacement" and use the Greek letter delta symbol (so Δx for 1-D as in Figure 1.2).

We relate time and position with velocity. Velocity is sometimes confused with speed, but it requires a direction as well as magnitude ("speed"). Even in 1-D velocity should include a direction since it could be moving to the left or to the right (if the 1-D line is arranged horizontally). The average velocity is simply the measure of the displacement divided by the change

Figure 1.1: Newtdog rolls Wormy in 1-D (©E. Diehl).

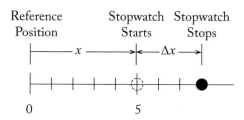

Figure 1.2: 1-D position and displacement.

in time during the motion, $v = \frac{\Delta x}{\Delta t}$. In Figure 1.2 you'd take Δx and divide it by the results of your stopwatch. You can find your average car speed if you divide your odometer change by the time it takes you to drive somewhere.

When we refer to velocity we most often mean the instantaneous velocity, $v = \frac{dx}{dt}$, or the time derivative of position as a function of time. This is the speed you read on your car's speedometer. Quite often we use "dot notation" for a time derivative, so velocity is $v = \dot{x}$.

Acceleration is the time rate change of velocity. The average acceleration is found from noting the velocity at two instances and dividing them by the time change between, $a = \frac{\Delta v}{\Delta t}$. The instantaneous acceleration is found when the time interval between these points gets smaller and smaller until $a = \frac{dv}{dt}$. Noting that velocity is the time derivative of position, we find acceleration is the second derivative of position with respect to time, $a = \frac{d^2 x}{dt^2}$. We can use dot notation for acceleration and write $a = \ddot{x}$. We can also relate acceleration to velocity and position by $a = v \frac{dv}{dx}$. Table 1.1 provides a summary of the above fundamental kinematic equations and relationships.

Table 1.1: Fundamental kinematic equations summary

	Generally	One Dimensional
Position at a particulat time	$\vec{\mathbf{r}}(t)$	Could be $\boldsymbol{x}$, $\boldsymbol{s}$, or $\boldsymbol{r}$
Average velocity ("speed" if only magnitude)	$\vec{\mathbf{v}} = \dfrac{\Delta\vec{\mathbf{r}}}{\Delta t}$	$v = \dfrac{\Delta x}{\Delta t}$
Average acceleration	$\vec{\mathbf{a}} = \dfrac{\Delta\vec{\mathbf{v}}}{\Delta t}$	$a = \dfrac{\Delta v}{\Delta t}$
Instantaneous velocity ("speed" if only magnitude)	$\vec{\mathbf{v}} = \dfrac{d\vec{\mathbf{r}}}{dt}$	$v = \dfrac{dx}{dt} = \dot{x}$
Instantaneous acceleration (related to velocity and time)	$\vec{\mathbf{a}} = \dfrac{d\vec{\mathbf{v}}}{dt}$	$a = \dfrac{dv}{dt} = \dot{v}$
Instantaneous acceleration (related to position and time)	$\vec{\mathbf{a}} = \dfrac{d^2\vec{\mathbf{r}}}{dt^2}$	$a = \dfrac{d^2x}{dt^2} = \ddot{x}$
Instantaneous acceleration (related to velocity and position)	$\vec{\mathbf{a}} = \vec{\mathbf{v}}\,\dfrac{d\vec{\mathbf{v}}}{d\vec{\mathbf{r}}}$	$a = v\,\dfrac{dv}{dx}$

We might ask why is this last acceleration relationship true? We can rearrange $v = \frac{dx}{dt}$ into $dt = \frac{dx}{v}$ and substitute it into $a = \frac{dv}{dt}$ to get $a = v\frac{dv}{dx}$.

These fundamental kinematic equations are in differential form, so we can arrive at a desired motion parameter by taking the derivative from position $\Rightarrow$ velocity $\Rightarrow$ acceleration or by integrating from acceleration $\Rightarrow$ velocity $\Rightarrow$ position. We'll demonstrate this with some examples.

There are actually more kinematic relations beyond acceleration since it isn't always constant and can change with time. So what is the time rate change of acceleration? The answer is the third time derivative of position, "jerk" $\left(\frac{d^3x}{dt^3}\right)$. This is a very descriptive term that we can perhaps envision. This begs the question, "what's the time rate change of jerk?" The answer to this and the next logical progression are "snap" (also "jounce") $\left(\frac{d^4x}{dt^4}\right)$, "crackle" $\left(\frac{d^5x}{dt^5}\right)$ and "pop" $\left(\frac{d^6x}{dt^6}\right)$. It would appear that someone with a sense of humor (and an affinity for a certain breakfast cereal) was involved in naming these. A practical application of this extended concept is the design of cam profiles. Cams are used for the timing of combustion engine valve opening and closing. The rise and fall of the cam follower depends on the cam profile/shape. That shape can be described by a lift displacement vs. time graph. The time derivatives of this displacement are graphs of velocity, acceleration, jerk, etc. Transitions occurring too quickly can result in theoretically infinite accelerations which aren't good for the equivalent forces on or by the cam follower, so the profile development is carried further to analyze jerk as well.

When applying these fundamental kinematic relations it is useful to visualize the motion and these relationships, and graphing is an excellent way to understand what's being described. Remember:

- kinematics is the "geometry of motion,"

- a derivative is the slope of a curve,

- integration gives the area under a curve, and

- inflection points in a curve are related to the maximums and minimums of the derivatives of the other curves.

We can check our work by graphing the results and looking for these properties. A simple example of derivatives is presented to serve as a review. You should replicate the table and graphs yourself in a spreadsheet.

Example 1.1
An object moves back and forth in a straight line as a function of time according to the equation $x(t) = 1 - \cos t$. What are the velocity and acceleration relationships as functions of time? Graph the position, velocity, and acceleration vs. time at 1 s intervals for 6 s.

$$v = \frac{dx}{dt} = \frac{d}{dt}[1 - \cos t] = \sin t$$
$$a = \frac{dv}{dt} = \frac{d}{dt}[\sin t] = \cos t.$$

We setup Table 1.2 below and find some position, velocity, and acceleration values at each time. The plot of these values (ignoring units) is shown in Figure 1.3. We can see this is a low resolution time increment but also note these sine and cosine functions have periods of $2\pi = 6.283$ so we've nearly graphed a full cycle.

Look for inflection points (where the curves cross the horizontal axis) and the corresponding maximums and minimums. For instance at about 3 s (3.142 s) position is at a maximum while velocity is zero. Acceleration at this point is also a maximum. We can conclude that the object is changing direction at this time, it has stopped moving, but is about to move. Here we address a possible misconception: just because an object has no velocity doesn't mean it has no acceleration. We often refer to this acceleration as "impending motion." Not only is there acceleration when something stops moving, it's potentially the point of maximum acceleration. We will encounter this situation in when applying Newton's 2nd Law (N2L) in Kinetics.

In Figure 1.4, we graph column 3 vs. column 2, column 4 vs. column 2, and column 4 vs. column 3. We can see some additional inter-relationships among the kinematic results.

Velocity vs. position and acceleration vs. velocity are nearly circles (they would be with more data points and symmetric axes). Acceleration vs. position is a straight line. You should try isolating a few points from Table 1.1 and find them on the graphs in Figure 1.4.

Table 1.2: Example 1.1 data

Column 1	Column 2	Column 3	Column4
t	$x(t)$	$v(t)$	$a(t)$
0	0.000	0.000	1.000
1	0.460	0.841	0.540
2	1.416	0.909	-0.416
3	1.990	0.141	-0.990
4	1.654	-0.757	-0.654
5	0.716	-0.959	0.284
6	0.040	-0.279	0.960

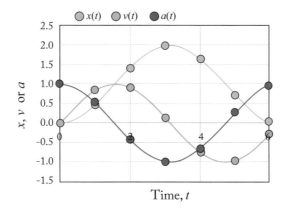

Figure 1.3: Example 1.1 results vs. time.

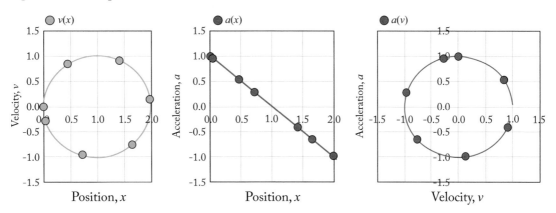

Figure 1.4: Example 1.1 results graphed kinematic relationships.

1.2   MOTION OF A PARTICLE VIA INTEGRATION

We found acceleration from the derivatives of position and velocity, but what if we wanted to know the position or velocity when the acceleration is known? We need to integrate, of course. There are a variety of scenarios we might need to consider.

Common cases of motion when the acceleration function is known:

1. The acceleration is a function of time, t. $a\left(t\right) = \frac{dv}{dt}$

   Rewrite: $dv = a\left(t\right)dt$

   Integrate: $\int_{v_0}^{v} dv = \int_{t_0}^{t} a(t)dt = v - v_0.$

2. The acceleration is a function of position, x. $a\left(x\right) = v\frac{dv}{dx}$

   Rewrite: $a\left(x\right)dx = v\,dv$

   Integrate: $\int_{x_0}^{x} a(x)dx = \int_{v_0}^{v} v\,dv = \frac{1}{2}\left(v^2 - v_0{}^2\right).$

3. The acceleration is a function of velocity, v. $a\left(v\right) = f\left(v\right)$

   Case 3.1: Time is desired, use $a(v) = \frac{dv}{dt}$

   Rewrite: $dt = \frac{dv}{a(v)}$

   Integrate: $\int_{t_0}^{t} dt = \int_{v_0}^{v} \frac{1}{a(v)}dv = t - t_0.$

   Case 3.2: Position is desired, use $a(v) = v\frac{dv}{dx}$

   Rewrite: $dx = \frac{v}{a\left(v\right)}dv$

   Integrate: $\int_{x_0}^{x} dx = \int_{v_0}^{v} \frac{v}{a(v)}dv = x - x_0,$

where:

   $t_0 =$ time value of start (often zero, but not necessarily),

   $x_0 =$ position at time t_0, and

   $v_0 =$ velocity at time t_0.

These were written so we can see the strategy here: identify what kind of function you have, reconfigure the fundamental kinematic equation, integrate. It is often more difficult to do this than it might seem from the above relations, as we'll see in the examples that follow.

An interesting way to think about time derivatives and integrals is derivatives "forecast" while integrals "remember." That is, a derivative provides a slope and therefore can point toward the possible next point. Integrals, however, provides the accumulation of what has already occurred, represented as the summation of the area under the curve. This concept is often discussed in control systems where derivative-type control elements will quicken the response (called "lead") by extrapolating the instantaneous slope to anticipate the next position, while integrator-type control elements will slow the response (called "lag") by accounting for what's already occurred (often the accumulation of error). There is much more to control systems than this, but one can imagine the importance of a stable control system in, for instance, an aircraft control stick. You'd like a quick response but don't want the plane to over-respond, so you need a balance between differential and integral control elements.

Example 1.2

A point is found to accelerate according to the relation $a = -1.5\sin(2t)$ m/s^2. At a reference starting time ($t = 0$), the position is $x = 0$ and the speed is $v = 1$ m/s. Find the position of the point when $t = 3.5$ s.

Acceleration is a function of time, therefore this is "Case 1" from fundamental kinematics. We use the known/initial values, but keep final as a variable, so a new function is written that describes the velocity as a function of time:

$$a = \frac{dv}{dt} \quad \Rightarrow \quad dv = a\,dt \quad \Rightarrow \quad \int dv = \int a\,dt$$

$$\int_1^v dv = \int_0^t [-1.5\sin(2t)]\,dt$$

$$v - 1 = \frac{1.5}{2}\cos(2t) - \frac{1.5}{2}$$

$$v(t) = 0.75\cos(2t) - 0.25.$$

Next, we use kinematic relation that relates velocity and time:

$$v = \frac{dx}{dt} \quad \Rightarrow \quad dx = v\,dt \quad \Rightarrow \quad \int dx = \int v\,dt$$

$$\int_0^x dx = \int_0^t [0.75\cos(2t) - 0.25]\,dt$$

$$x - 0 = \frac{0.75}{2}\sin(2t) - 0.25t$$

$$x(t) = 0.375\sin(2t) - 0.25t.$$

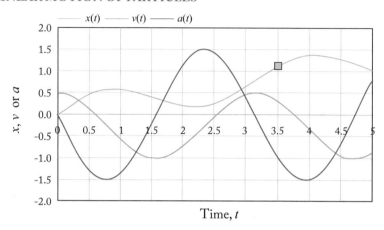

Figure 1.5: Example 1.2 results.

This describes position as a function of time, which was our goal since we were asked for the position at a particular time:

$$x(3.5) = 0.375\sin(2(3.5)) - 0.25(3.5) = \boxed{1.121\text{ m}}.$$

Remember to switch your calculator to radians, which we always assume is the unit used in sine, cosine, and tangent functions (unless told otherwise).

Graphing the functions (Figure 1.5) helps get a better understanding of what's really going.

Example 1.3 The acceleration of an object is approximated by $a = 12 + 0.05x$, where a and x are expressed in m/s^2 and m, respectively. At a set reference point, $t = 0$ and $x = 0$, the velocity is $v = 10$ m/s. Determine the time when the object is in position $x = 40$ m.

Acceleration is a function of position, therefore this is "Case 2" from fundamental kinematics:

$$a = v\frac{dv}{dx} \quad\Rightarrow\quad v\,dv = a\,dx \quad\Rightarrow\quad \int v\,dv = \int a\,dx$$

$$\int_{10}^{v} v\,dv = \int_{0}^{x} [12 + 0.05x]dx$$

$$\frac{1}{2}v^2 - \frac{1}{2}(10)^2 = 12x + \frac{0.05}{2}x^2$$

$$v(x) = \sqrt{0.05x^2 + 24x + 100}.$$

We use the known/initial values but keep final as a variable so a new function is written that describes the velocity as a function of position:

$$v = \frac{dx}{dt} \quad \Rightarrow \quad dt = \frac{dx}{v} \quad \Rightarrow \quad \int dt = \int \frac{1}{v}\, dx$$

$$\int_0^t dt = \int_0^x \left[\frac{1}{\sqrt{0.05x^2 + 24x + 100}} \right] dx.$$

This is not a common integral form, so we'll need to look it up from an integral table. From `http://integral-table.com` we find the correct form:

$$\int \left[\frac{1}{\sqrt{ax^2 + bx + c}} \right] dx = \frac{1}{\sqrt{a}} \ln \left[2ax + b + 2\sqrt{a\left(ax^2 + bx + c\right)} \right]$$

$$t = \frac{1}{\sqrt{0.05}} \ln \left[2\,(0.05)\,x + (24) + 2\sqrt{(0.05)\left((0.05)\,x^2 + (24)\,x + 100\right)} \right] \Bigg|_0^x$$

$$t(x) = (4.472) \ln \left[(0.1)\,x + (24) + 2\sqrt{(0.0025)\,x^2 + (1.2)\,x + 5} \right] - 14.98.$$

This describes time as a function of position, which was our goal since we were asked for the time at a particular position:

$$t(x) = (4.472) \ln \left| (0.1)(40) + (24) + 2\sqrt{(0.0025)(40)^2 + (1.2)(40) + 5} \right| - 14.98$$

$$= \boxed{1.854 \text{ s}}.$$

Since we can find time, velocity and acceleration values for increments of position, we generate a similar plot (Figure 1.6) as the previous examples.

Example 1.4

The device in the figure shown (Figure 1.7) is referred to as a "slider-crank" in mechanisms and often represents the piston (A), connecting rod (AB), and crank shaft (BC) of an internal combustion engine among other basic machines. We will use them extensively in rigid body dynamics problems, but for particle kinematics we'll limit this to the movement of the piston which we assume we know. For a limited range of motion (not the entire cycle), the acceleration of A can be described by the relation $a = 0.5\sqrt{9 - v^2}$, where a and v are expressed in in/s^2 and in/s, respectively. The piston starts from rest ($t = 0$ and $v = 0$) at $x = 7$ in. Determine the position when $t = 2$ s.

Acceleration is a function of velocity, therefore this is "Case 3." Let's try "Case 3.1" to see what happens:

$$a = \frac{dv}{dt} \quad \Rightarrow \quad dt = \frac{dv}{a} \quad \Rightarrow \quad \int_{t_0}^t dt = \int_{v_0}^v \frac{1}{a}\, dv$$

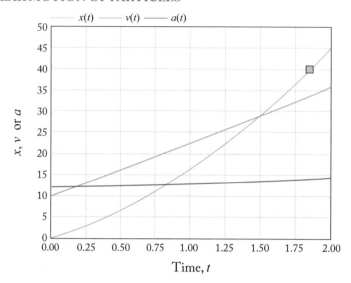

Figure 1.6: **Example** 1.3 results.

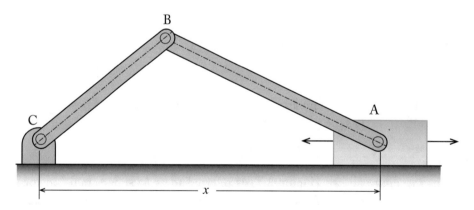

Figure 1.7: Slider crank.

$$\int_0^t dt = \int_0^v \frac{1}{0.5\sqrt{9-v^2}} dv.$$

From `http://integral-table.com` we find the surprising form: $\int \left[\frac{1}{\sqrt{a^2-x^2}} \right] dx = \sin^{-1} \frac{x}{a}$

$$t = \left(\frac{1}{0.5} \right) \sin^{-1} \left(\frac{v}{\sqrt{9}} \right) = (2) \sin^{-1} \left(\frac{v}{3} \right).$$

We now can rearrange to find velocity as a function of time:

$$v(t) = 3 \sin\left(\frac{t}{2}\right)$$

$$v(2) = 3 \sin\left(\frac{2}{2}\right) = 2.524 \text{ in/s.}$$

Use the definition of velocity to get the position. (Note: We might also find the position as a function of velocity from the acceleration function using $a\,(v) = v\frac{dv}{dx}$, which is Case 3.2. But that is more complicated than is necessary.)

$$v = \frac{dx}{dt} \quad \Rightarrow \quad dx = v\,dt \quad \Rightarrow \quad \int dx = \int v\,dt$$

$$\int_{x_0}^{x} dx = \int_{0}^{t} \left[3 \sin\left(\frac{t}{2}\right)\right] dt$$

$$x - x_0 = -(2)(3) \cos\left(\frac{t}{2}\right)\Big|_{0}^{t}$$

$$x(t) = x_0 - (6) \cos\left(\frac{t}{2}\right) + (6) = (7) - (6) \cos\left(\frac{t}{2}\right) + (6)$$

$$= (13) - (6) \cos\left(\frac{t}{2}\right)$$

$$x(t) = (13) - (6) \cos\left(\frac{t}{2}\right).$$

This describes position as function of time which allows us to find the desired result:

$$x(2) = (13) - (6) \cos\left(\frac{2}{2}\right) = \boxed{9.757 \text{ in}}\,.$$

1.3   PARTING THOUGHTS AFTER FIRST CLASS

This first topic is a great place to start the course, because it introduces the students' Dynamics Dilemma: "why are the homework problems so much harder than the examples" or even worse yet, "why are the exams so much harder than the homework?" Dynamics makes us apply concepts rather than go through canned procedures. While there are some methodical approaches you should learn throughout this book, there are many MANY situations that are frustratingly more complex.

Often students approach math and science like this: find the right equation, enter values into it and get an answer: "plug-and-chug." But real problems, the kind engineers are paid a lot to solve, often aren't like that, they haven't been solved before, so there is no step-by-step procedure. For this reason the struggle you may have in Dynamics will pay off as a valuable learning experience, allowing you to develop those desired problem solving skills, make you dig deep and think hard. The only way to get better at solving difficult problems is to solve increasingly more difficult problems. Similarly, the only way to measure students' mastery of problem solving skills is to test them with problems they've not yet solved. New problems almost always seem more difficult than they are. Evidence of this is that their solutions typically seem somewhat obvious in hindsight. Your professor will likely test your problem solving skills this way. Don't worry, your Course Companion is here to help you build up those problem solving skills.

Book 1 - Class 2

https://www.youtube.com/watch?v=4WZUiC9W13k

CLASS 2

Kinematics Special Cases: One-Dimensional Relative Motion and Dependent Motion

B.L.U.F. (Bottom Line Up Front)

- Kinematic Special Cases:

    - Constant Velocity

    $$\vec{r} = \vec{r}_0 + \vec{v}_c t \qquad \text{In one dimension:} \quad x = x_0 + v_c t$$

    - Constant Acceleration

    $$\vec{v} = \vec{v}_0 + \vec{v}_c t \qquad \text{In one dimension:} \ v = v_0 + a_c t$$

    $$\vec{r} = \vec{r}_0 + \vec{v}_0 t + \tfrac{1}{2} \vec{v}_c t^2 \qquad \text{In one dimension:} \ x = x_0 + v_0 t + \tfrac{1}{2} a_c t^2$$

    $$\vec{v}^2 = \vec{v}_0^2 + 2 \vec{a}_c (\vec{r} - \vec{r}_0) \qquad \text{In one dimension:} \ v^2 = {v_0}^2 + 2 a_c (x - x_0)$$

- Relative Motion (one-dimensional)

    $$\vec{r}_B = \vec{r}_A + \vec{r}_{B/A} \qquad \text{In one dimension:} \quad x_B = x_A + x_{B/A}$$

    $$\vec{v}_B = \vec{v}_A + \vec{v}_{B/A} \qquad \text{In one dimension:} \quad v_B = v_A + v_{B/A}$$

    $$\vec{v}_B = \vec{v}_A + \vec{v}_{B/A} \qquad \text{In one dimension:} \quad a_B = a_A + a_{B/A}$$

- Dependent Motion: Pulley Problems and the "conservation of rope."

2.1   PARTICLE KINEMATICS SPECIAL CASES

The fundamental kinematics equations discussed in Class 1 are general and require that the velocity or acceleration function is known. The special cases described here occur when either the velocity or acceleration is constant. Only the single dimension versions are derived below using x as a generic position parameter for convenience, but the vector versions are also true.

Constant velocity $a(t) = 0, v(t) = constant = v_c$

$$v = \frac{dx}{dt} \quad \Rightarrow \quad dx = v_c dt \quad \Rightarrow \quad \int dx = v_c \int dt$$

$$\int_{x_0}^{x} dx = v_c \int_{0}^{t} dt \Rightarrow \quad \boxed{x = x_0 + v_c t} \ .$$

Constant acceleration $a(t) = constant = a_c$

In terms of time: $a = \frac{dv}{dt} \quad \Rightarrow \quad dv = a_c dt \quad \Rightarrow \quad \int dv = a_c \int dt$

$$\int_{v_0}^{v} dv = a_c \int_{0}^{t} dt \quad \Rightarrow \quad \boxed{v(t) = v_0 + a_c t} \ .$$

$$v = \frac{dx}{dt} \quad \Rightarrow \quad dx = v(t)\, dt \quad \Rightarrow \quad \int dx = \int v(t) dt$$

$$\int_{x_0}^{x} dx = \int_{0}^{t} [v_0 + a_c t]\, dt \quad \Rightarrow \quad \boxed{x(t) = x_0 + v_0 t + \frac{1}{2} a_c t^2} \ .$$

In terms of position:

$$a = v\frac{dv}{dx} \quad \Rightarrow \quad v dv = a_c dx \quad \Rightarrow \quad \int v dv = a_c \int dx$$

$$\int_{v_0}^{v} v dv = a_c \int_{x_0}^{x} dx \quad \Rightarrow \quad \frac{1}{2}\left(v^2 - v_0{}^2\right) = a_c (x - x_0)$$

$$\Rightarrow \quad \boxed{v(x)^2 = v_0{}^2 + 2a_c (x - x_0)} \ .$$

Be sure to use these equations ONLY for constant cases.

Example 2.1

A bottle rocket is set off and travels $y = 300$ ft straight up when the fuel runs out but keeps traveling upward (Figures 2.1 and 2.2). Using a stop watch, it's found that it lands on the ground $\Delta t = 10$ s after the fuel runs out. Assume that the acceleration downward is equal to gravity ($a_c = g = 32.2$ ft/s^2) and determines the maximum height it reached. *Note:* This is technically a "projectile motion" problem, except it is only in one dimension and therefore rectilinear. Projectile motion problems are covered in more detail in Class 3.

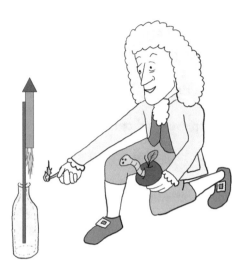

Figure 2.1: Newtdog fires off a bottle rocket (©E.Diehl).

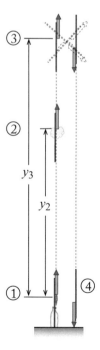

Figure 2.2: Bottle rocket stages of Example 2.1.

As with all Dynamics problems, you should visualize what's taking place, think about what's being asked, identify possible ways to approach getting a solution and try some. The first step is often the most important and difficult, so a good approach is to first make a sketch. In Figure 2.2, we sketch the events using the center of the bottle rocket as a reference to treat it as a particle. The stages are:

①–②   It goes up due to the fuel. (Note this is the kinetics portion.)

②–③   It keeps going up even without fuel… until it stops and $v_3 = 0$.

③–④   It comes crashing back to Earth.

We know this is a constant acceleration problem and therefore the appropriate equations are all true between any two points. Using y as the variable and points ② and ④, find the velocity at ②, since we know the time between ② and ④ is $\Delta t = 10$ s.

$$y_4 = y_2 + v_2 t + \frac{1}{2} a_c t^2,$$

where:

$y_4 = 0$ when it hits the ground

$y_2 = 300$ ft when the time starts

$t = 10$ s

$a_c = g = -32.2$ ft/s^2

(negative because upwards is positive)

$$0 = (300) + v_2 (10) + \frac{1}{2}(-32.2)(10)^2$$
$$v_2 = 131.0 \text{ ft/s}.$$

Now we can use another piece of information the sketch helps to recognize: the velocity at the peak/zenith is zero. We can find the position where this happens:

$$v_3{}^2 = v_2{}^2 + 2a_c (y_3 - y_2)$$
$$(0)^2 = (131.0)^2 + 2(-32.2)(y_3 - (300))$$
$$\boxed{y_3 = 566.5 \text{ ft}}.$$

This was probably not an obvious approach at first reading of the problem. Memorizing this approach is only useful if this exact problem were asked again, instead we want to think through possible approaches to find a path to a solution.

Figure 2.3: Newtdog and Wormy demonstrate relative motion (©E. Diehl).

2.2   PARTICLE KINEMATICS RELATIVE MOTION (1-D)

We are all familiar with relative motion as experienced in cars on the highway. When a speeder comes zipping by us while we're going the speed limit, we get a sense of their actual speed by their apparent speed (we can reference our past experiences with driving by stationary things at known speeds) added to our known speed (the speed limit, of course). This is their relative speed. In Figure 2.3, Newtdog is passing Wormy since he's traveling faster. Wormy perceives Newtdog to being going by him at the relative velocity.

We can formalize this and say "with respect to" for the relative speed. If we are A, and the speeder is B, we can say the relative speed is "B with respect to A" or as an equation with subscripts: $v_{B/A} = v_B - v_A$. If we assign directions to it (positive or negative in one dimension), this relationship still holds true, but we note that if we swap which is the reference, the sign will also change: $v_{A/B} = v_A - v_B$, so $v_{B/A} = -v_{A/B}$.

This can also be done with distances as shown in Figure 2.4. Note the origin, "O", can also be used as a reference, so variables can be written as $x_A = x_{A/O}$ and $x_B = x_{B/O}$ resulting in the equation $x_{B/A} = x_{B/O} - x_{A/O}$. This reference is usually omitted since it's understood there is always an origin. A typical application is to find an unknown value, like the position of B, by taking a known position, here we use A, and adding the relative position, so we have "B with respect to A" = $x_{B/A}$, defined as $x_{B/A} = x_B - x_A$ and rewritten as:

$$\boxed{x_B = x_A + x_{B/A}}.$$

The time rate change of the position equals the velocity $\left(v = \frac{dx}{dt}\right)$, so:

$$\frac{d\left(x_B\right)}{dt} = \frac{d\left(x_A\right)}{dt} + \frac{d\left(x_{B/A}\right)}{dt} \quad \Rightarrow \quad \boxed{v_B = v_A + v_{B/A}}.$$

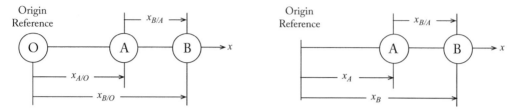

Figure 2.4: Relative position.

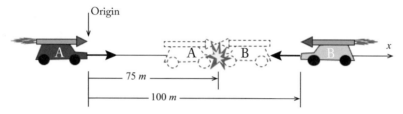

Figure 2.5: Rocket cars collide in Example 2.2.

Likewise the time rate change of the velocity equals acceleration $\left(a = \frac{dv}{dt}\right)$, so:

$$\frac{d\left(v_B\right)}{dt} = \frac{d\left(v_A\right)}{dt} + \frac{d\left(v_{B/A}\right)}{dt} \quad \Rightarrow \quad \boxed{a_B = a_A + a_{B/A}} \, .$$

This is shown for one dimension, but holds true for generalized vectors which will be covered in Class 3:

$$\vec{r}_B = \vec{r}_A + \vec{r}_{B/A} \qquad \vec{v}_B = \vec{v}_A + \vec{v}_{B/A} \qquad \vec{a}_B = \vec{a}_A + \vec{a}_{B/A}.$$

This seemingly simple concept will be used extensively in rigid body motion to relate points on a rigid body to one another.

Example 2.2
Two toy rocket cars capable of a different constant acceleration are aimed toward each other starting 100 m apart. Released at the same time, toy rocket car A collides with B at 450 m/s after it travels 75 m (Figure 2.5). Find the relative velocity and relative acceleration of car B with respect to A.

Treat rocket car A's starting point as the reference origin and apply what we know about its motion to find its acceleration and the time when it reaches the collision point:

$$v_A = (v_A)_0 + 2a_A\left(x_A - (x_A)_0\right)$$

$$(900) = (0) + 2a_A\left((75) - (0)\right) \qquad a_A = 6.000 \text{ m/s}^2 \quad \rightarrow$$

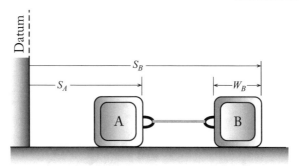

Figure 2.6: Blocks linked by rope and dependent motion in one plane.

$$x_A = (x_A)_0 + (v_A)_0 t + \frac{1}{2}a_A t^2$$

$$(75) = (0) + (0)t + \frac{1}{2}(6.000)t^2 \qquad t = 5.000 \text{ s}$$

Using this time and the remaining distance, find rocket car B's acceleration:

$$x_B = (x_B)_0 + (v_B)_0 t + \frac{1}{2}a_B t^2$$

$$(75) = (100) + (0)(5.000) + \frac{1}{2}a_B(5.000)^2 \qquad a_B = -2.000 \text{ m/s}^2 = 2.000 \text{ m/s}^2 \;\leftarrow$$

$$v_B = (v_B)_0 + a_B t \quad v_B = (0) + (-2.000)(5.000) = -10.00 \text{ m/s} = 10.00 \text{ m/s} \;\leftarrow$$

$$v_{B/A} = v_B - v_A = (-10.00) - (900) = -910.0 \text{ m/s} = \boxed{910.0 \text{ m/s} \;\leftarrow}$$

$$a_{B/A} = a_B - a_A = (-2.000) - (6.000) = -8.000 \text{ m/s}^2 = \boxed{8.000 \text{ m/s}^2 \;\leftarrow}.$$

This is a very simple problem to demonstrate the relative motion concept. Obviously, we could face more complicated problems reminiscent of those algebra word problems involving trains we hated in high school. Here though we can use the simple relative motion equations to organize the problems, along with fundamental kinematics to come up with a path to a solution.

2.3   DEPENDENT MOTION

Situations arise where the motion of particles are linked to each other, "depend" on each other. The problems we'll solve here are all "pulley problems" where the particles are linked by rope. The position, velocity and acceleration of blocks A and B in Figure 2.6 are dependent on each other

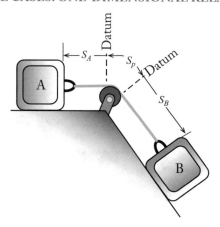

Figure 2.7: Newtdog and Wormy depending on each other with multiple ropes.

Figure 2.8: Dependent motion blocks in different planes (©E. Diehl).

and as long as the rope can't stretch are equal. Other examples of dependent motion might be the height of a shadow as an object and/or the light source moves. Another instance of dependent motion we'll cover is rigid body motion where the motion of points on the body are dependent.

The objective is to find equations relating the particles motion which can be the extra equation(s) we need to solve for the unknowns. For pulleys we can think of this as "the conservation of rope." If the rope can't extend, its overall length remains constant. The general procedure we follow is to establish coordinates for reference points, add up segment lengths, and take derivatives of the position relations to find how the velocities and accelerations relate (Figures 2.7 and 2.8). Even though we might have multiple directions of motion, these problems are still 1 DOF since the motion can only be positive and negative along a line for each particle. What follows is a more detailed procedure.

1. Create reference locations (aka "datum," $+/-$), define position coordinates to each point that moves, establishing a coordinate system.

2. Write an expression (equation) for the *length of the cable* in terms of position coordinates. Make use of constant distances (like length of the block or the arc of rope around a pulley that doesn't change) to lump together with constant cable length into an overall constant. The following references Figure 2.7:

$$s_A + s_p + s_B = length$$

$$s_A + s_B = length - s_p = constant.$$

3. Take time derivatives of the expressions to get relationships in terms of velocity and acceleration:

$$\frac{d}{dt}[s_A + s_B] = \frac{d}{dt}[constant]$$

$$v_A + v_B = 0 \qquad v_A = -v_B$$

$$a_A + a_B = 0 \qquad a_A = -a_B.$$

4. Be careful to consider positive and negative results as referencing motion relative to the datums (remember we are still treating motion as one-dimensional). Reference this simple example to help interpret the results of future problems.

5. There are equations for each individual rope in more complex pulley arrangements.

Example 2.3

Find the velocity and acceleration relations between blocks A and B (Figure 2.9).

A methodical approach is to trace the path of each rope and find the length of every segment in terms, which we'll do for this example. But we'll soon see that many of the constant portions can be ignored because they go away when differentiated:

$$\underbrace{(s_A - h)}_{①} + l_{p1} + \underbrace{(s_B - l_2 - h)}_{②} + l_{p2} + \underbrace{(s_B - l_2)}_{③} = length$$

$$s_A + 2s_B = length + 2h - 2l_2 - l_{p1} - l_{p2} = constant$$

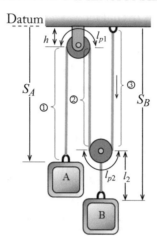

Figure 2.9: Example 2.3.

$$\frac{d}{dt}[s_A + 2s_B] = \frac{d}{dt}[constant]$$

$$v_A + 2v_B = 0 \qquad \boxed{v_A = -2v_B}$$

$$a_A + 2a_B = 0 \qquad \boxed{a_A = -2a_B}.$$

We could have observed that there are three varying length segments, one of A and two of B, so: $s_A + 2s_B = constant$. We'll take this short-cut for future problems.

Example 2.4
Find the velocity and acceleration of C in terms of blocks A and B (Figure 2.10):

$$s_A + 2s_B + s_C = constant$$

$$\frac{d}{dt}[s_A + 2s_B + s_C] = \frac{d}{dt}[constant]$$

$$v_A + 2v_B + v_C = 0 \qquad \boxed{v_C = -2v_B - v_A}$$

$$a_A + 2a_B + a_C = 0 \qquad \boxed{a_C = -2a_B - a_A}.$$

Note there are three "degrees of freedom" here, so you need to know motion of two points to determine the motion of the third.

Example 2.5
Find the velocity relationship between blocks A and B (Figure 2.11).

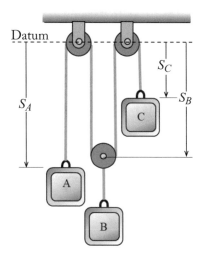

Figure 2.10: **Example** 2.4.

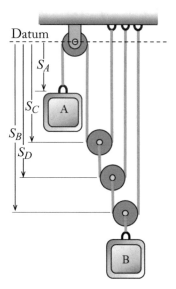

Figure 2.11: **Example** 2.5.

This example demonstrates when there is an equation for each rope.

Rope 1: $s_A + 2s_C = l_1$

$$\frac{d}{dt}[s_A + 2s_C] = \frac{d}{dt}[l_1]$$

$$v_A + 2v_C = 0$$

$$v_C = -\frac{1}{2}v_A.$$

Rope 2: $\quad s_D + s_D - s_C = l_2$

$$\frac{d}{dt}[2s_D - s_C] = \frac{d}{dt}[l_2]$$

$$2v_D - v_C = 0$$

$$v_D = \frac{1}{2}v_C = -\frac{1}{4}v_A.$$

Rope 3: $\quad s_B + s_B - s_D = l_3$

$$\frac{d}{dt}[2s_B - s_D] = \frac{d}{dt}[l_3]$$

$$2v_B - v_D = 0$$

$$v_B = \frac{1}{2}v_D = -\frac{1}{8}v_A \quad \boxed{v_B = -\frac{1}{8}v_A} .$$

Example 2.6

Block A starts from rest and moves downward with a constant acceleration. Knowing that after 5 s the velocity of Newtdog relative to Wormy is 8 ft/s downward, determine the acceleration of block A (Figure 2.12).

$$v_{N/W} = 8 \text{ ft/s } \downarrow = -8 \text{ ft/s}$$

$$v_{N/W} = (v_{N/W})_0 + a_{N/W}t$$

$$(-8) = (0) + a_{N/W}(5)$$

$$a_{N/W} = -1.600 \text{ ft/s}^2 = 1.600 \text{ ft/s}^2 \downarrow$$

Rope 1:

$$s_N + 2s_A = l_1 \quad \frac{d}{dt}[s_N + 2s_A] = \frac{d}{dt}[l_1]$$

$$v_N + 2v_A = 0 \quad a_N + 2a_A = 0$$

$$a_N = -2a_A \quad ①$$

Figure 2.12: Example 2.6, repeat of Figure 2.8 (©E. Diehl).

Rope 2:

$$3s_A + s_W = l_2 \quad \frac{d}{dt}[3s_A + s_W] = \frac{d}{dt}[l_2]$$
$$3v_A + v_W = 0 \quad 3a_A + a_W = 0$$
$$a_W = -3a_A \quad ②$$
$$a_{N/W} = a_N - a_W \quad ③$$

Combine ①, ②, and ③

$$(-1.600) = -2a_A - (-3a_A) \quad \boxed{a_A = 1.600 \text{ ft/s}^2 \ \uparrow}.$$

An important Dynamics problem solving skill is keeping track of the direction of motion. While using a coordinate system and being disciplined with your sign conventions is very key, using logic to double check the results make sense is also critically important.

In this class we've looked at special cases where the fundamental kinematic relationships are already integrated. We'll use these again in the next class using more than one dimension, specifically for projectile motion. We also introduced simple relative motion and dependent motion. When solving complex problems with multiple degrees of freedom, these are often useful as additional equations when the application of other principles yields fewer equations than unknowns.

https://www.youtube.com/watch?v=0T8lhwqZ3C8

Curvilinear Motion of Particles (Rectangular Coordinates): Projectile Motion and Vector Relative Motion

B.L.U.F. (Bottom Line Up Front)

- Review of vector addition: add components or use triangles.

- Derivatives of vectors: must take derivative of unit vectors if they change.

- Curvilinear Motion of Particles: rectangular coordinates have simple coefficients.

$$\vec{r} = x\hat{i} + y\hat{j} \qquad \vec{v} = v_x\hat{i} + v_y\hat{j} \qquad \vec{a} = a_x\hat{i} + a_y\hat{j}$$

- Projectile Motion: horizontal constant velocity, vertical constant acceleration.

$$x = x_0 + (v_0)_x t \qquad y = y_0 + (v_0)_y t - \frac{1}{2}gt^2 \qquad v_y = (v_0)_y - gt$$

- Relative Motion can use a reference vector to define position and motion of a point.

$$\vec{r}_B = \vec{r}_A + \vec{r}_{B/A} \qquad \vec{v}_B = \vec{v}_A + \vec{v}_{B/A} \qquad \vec{a}_B = \vec{a}_A + \vec{a}_{B/A}$$

3.1   VECTOR ADDITION REVIEW

This section reviews how vectors are defined and added together before we use them for kinematics. We'll restrict ourselves to two dimensions for simplicity and start with rectangular coordinates by default, but recognize these concepts also extend to other coordinate systems. Skip this section if you're already comfortable with it and/or will come back and use it for reference while doing problems.

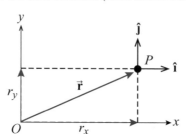

Figure 3.1: Position vector components.

To define the location of a particle (P) we use a position vector, $\vec{r}$, by establishing an origin (O), a reference coordinate system with unit vectors providing direction, and component coefficients for magnitudes along each unit vector. The rectangular coordinates position vector in two-dimensions shown in Figure 3.1 is written $\vec{r} = r_x\hat{i} + r_y\hat{j}$ (or is often written as $\vec{r} = x\hat{i} + y\hat{j}$), where the component coefficients are also called just "components" or just "coordinates" to plot the particle position. This could also be written as $\vec{r} = r_x\hat{e}_x + r_y\hat{e}_y$ if we were to be consistent.

The unit vectors can be thought of as attached to the point. Each unit vector is parallel to the coordinate axis. This orientation doesn't change in rectangular coordinates but can rotate with the particle motion in other coordinate systems. We denote rectangular coordinate unit vectors as $\hat{i}, \hat{j}$, and $\hat{k}$ because that's the standard convention. Unit vectors are "orthogonal," which means perpendicular, and follow the "right-hand rule" where the thumb and index finger form the x and y coordinates, respectively, when making an "L" shape and a third coordinate, z, follows the direction of the middle finger when it is extended away from the palm. In Figure 3.2, Newtdog is demonstrating the right-hand-rule. This convention is also used for defining other 3-D coordinate systems and for defining positive rotation directions.

Vector addition (for instance: $\vec{f} = \vec{d} + \vec{e}$) can be done by either adding up the components or by positioning representative vectors into triangles and working through the geometry. Figure 3.3 depicts these vectors as arrows and shows how they add to form a new vector.

Addition using components: Break each vector up into components and add them to generate the new vector. Smaller arrows in Figure 3.3 show the components and how they add to form the new vector:

$$\vec{f} = (d_x + e_x)\hat{i} + (d_y + e_y)\hat{j}.$$

Addition using triangles: Vector addition can be visualized by the triangle, as shown in Figure 3.3. One convention is to place vectors nose to tail to add together to create new vector which starts at the tail of the first vector and ends at the nose of the second. A way to remember this is to think of the addition is "nose-tail" and the result is "tail-tail" and "nose-nose."

Figure 3.2: Newtdog and the right-hand rule (©E. Diehl).

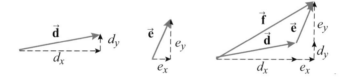

Figure 3.3: Vector components forming vector triangles.

The geometry of the vector triangle can be used to find the magnitude and direction of the new vector. The Law of Sines (LoS) and Law of Cosines (LoC) are often useful to find the angles (upper case letters for the angle opposite from the referenced vector) and lengths (lowercase letters representing the vector magnitude) (Figure 3.4).

Recall:

$$\text{LoS:} \quad \frac{d}{\sin D} = \frac{e}{\sin E} = \frac{f}{\sin F}$$
$$\text{LoC:} \quad f^2 = d^2 + e^2 - 2de \cos F.$$

There are a variety of scenarios of known magnitudes and/or angles we might encounter where we can use LoS and/or LoC to find the unknowns. For instance, if we know the magnitude and direction of $\vec{d}$ and $\vec{e}$, we use the lengths of d and e and angle F. To find the length of $\vec{f}$ from LoC and then the remaining angles from LoS. Sometimes we may know $\vec{f}$ and

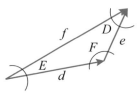

Figure 3.4: Vector triangle to use LoS and LoC.

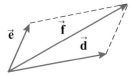

Figure 3.5: Vector addition using parallelogram.

want to get find either $\vec{d}$ or $\vec{e}$ and therefore need to use the other two versions of LoC: $d^2 = e^2 + f^2 - 2ef \cos D$, and $e^2 = d^2 + f^2 - 2df \cos E$.

You can also use a parallelogram (Figure 3.5) to visualize the new vector by putting the two known vectors tail to tail to find the new vector, but the geometry from this can't be easily used to find the new vector. Still, the parallelogram method is useful to visualize the new vector and confirm the results from the other methods. You should use multiple methods when time permits to catch mistakes and gain confidence in your work.

3.2   DERIVATIVE OF VECTORS REVIEW

To explain the reason for (and need for) some of the complications that arise with non-rectangular coordinate systems, we must do a bit of generalized vector derivatives. The bottom line up front of this topic is: *the product rule comes into play when the orientation of the unit vectors change with respect to time.* Skip this section if you're already comfortable with the concepts presented.

Derivative of a scalar function times a vector function: $f\vec{P}$
Where

$$f = f(t) = \quad \text{some scalar function that changes with time}$$

$$\vec{P} = \vec{P}(t) = \quad \text{some vector function that changes with time}$$

$$\frac{d(f\vec{P})}{dt} = \frac{df}{dt}\vec{P} + f\frac{d\vec{P}}{dt}.$$

Derivative of a scalar product: $\frac{d(\vec{P}\cdot\vec{Q})}{dt}$

Where $\vec{\mathbf{Q}} = \vec{\mathbf{Q}}(t)$ = some other vector function that changes with time

$$\frac{d\left(\vec{\mathbf{P}} \cdot \vec{\mathbf{Q}}\right)}{dt} = \frac{d\vec{\mathbf{P}}}{dt} \cdot \vec{\mathbf{Q}} + \vec{\mathbf{P}} \cdot \frac{d\vec{\mathbf{Q}}}{dt}.$$

Derivative of a vector product: $\frac{d(\vec{\mathbf{P}} \times \vec{\mathbf{Q}})}{dt}$

$$\frac{d\left(\vec{\mathbf{P}} \times \vec{\mathbf{Q}}\right)}{dt} = \frac{d\vec{\mathbf{P}}}{dt} \times \vec{\mathbf{Q}} + \vec{\mathbf{P}} \times \frac{d\vec{\mathbf{Q}}}{dt}.$$

Since within a vector the components are scalars and the unit vectors are vectors, written as $\vec{\mathbf{P}} = P_1\hat{\mathbf{e}}_1 + P_2\hat{\mathbf{e}}_2$ or $\vec{\mathbf{P}}(t) = P_1(t)\,\hat{\mathbf{e}}_1(t) + P_1(t)\,\hat{\mathbf{e}}_2(t)$ (where $\hat{\mathbf{e}}_1$ and $\hat{\mathbf{e}}_2$ are generalized unit vectors), the derivative of any vector itself is:

$$\frac{d\vec{\mathbf{P}}}{dt} = \frac{dP_1}{dt}\hat{\mathbf{e}}_1 + P_1\frac{d\hat{\mathbf{e}}_1}{dt} + \frac{dP_2}{dt}\hat{\mathbf{e}}_2 + P_2\frac{d\hat{\mathbf{e}}_2}{dt}.$$

For rectangular coordinates $(\vec{\mathbf{P}} = P_x\hat{\imath} + P_y\hat{\jmath})$ this becomes:

$$\frac{d\vec{\mathbf{P}}}{dt} = \frac{dP_x}{dt}\hat{\imath} + P_x\frac{d\hat{\imath}}{dt} + \frac{dP_y}{dt}\hat{\jmath} + P_y\frac{d\hat{\jmath}}{dt}.$$

But since the unit vectors don't change in rectangular coordinates, $\frac{d\hat{\imath}}{dt} = 0$ and $\frac{d\hat{\jmath}}{dt} = 0$, this becomes a non-issue resulting in:

$$\frac{d\vec{\mathbf{P}}}{dt} = \frac{dP_x}{dt}\hat{\imath} + \frac{dP_y}{dt}\hat{\jmath}.$$

So why bother describing this? Because the unit vectors in other coordinate systems do change with time and will require us to consider the product rule when taking a time derivative. This results in components with multiple parts.

3.3   CURVILINEAR MOTION OF PARTICLES

3.3.1   VELOCITY VECTOR FROM POSITION VECTOR

Recall from Class 1, the instantaneous velocity is defined as $\vec{\mathbf{v}} = \frac{d\vec{\mathbf{r}}}{dt}$. To visualize this, imagine a point P and follow it along the path of its motion to a second point at a later time P'. The position vectors for these two points are $\vec{\mathbf{r}}$ and $\vec{\mathbf{r}}'$. Figure 3.6 illustrates this motion.

We recall the definition of a derivative (the limit of the changing function as the reference parameter approaches zero) as it applies to velocity:

$$\vec{\mathbf{v}} = \lim_{\Delta t \to 0} \frac{\Delta \vec{\mathbf{r}}}{\Delta t} = \frac{d\vec{\mathbf{r}}}{dt}.$$

Compare Figures 3.6 (a)–(c), and note that as Δt shrinks to dt, the velocity vector becomes tangent to the path. This result provides us with a very important conclusion we should remember: *velocity is always tangent to the path.*

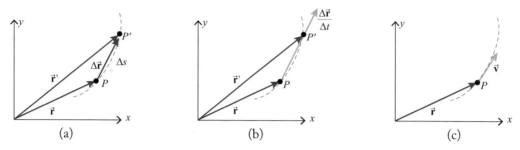

Figure 3.6: Position vectors used to explain velocity being tangent to path.

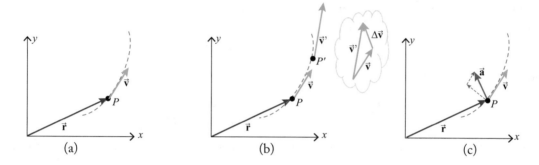

Figure 3.7: Velocity vectors used to explain acceleration components.

3.3.2 ACCELERATION VECTOR FROM VELOCITY VECTOR

A similar treatment for acceleration is performed with $\vec{a} = \lim_{\Delta t \to 0} \frac{\Delta \vec{v}}{\Delta t} = \frac{d\vec{v}}{dt}$ in Figure 3.7.

The velocity vectors attached to points P and P' begin to merge as Δt shrinks to dt. The triangle shown in the thought bubble illustrates this important concept: *the change in velocity direction is NOT tangent to the path*. This is because the change of direction also contributes to the acceleration. We'll see in the next class (when using other coordinates) exactly how acceleration breaks down. The important concept: *acceleration is NOT tangent to the path*.

3.3.3 VELOCITY AND ACCELERATION IN RECTANGULAR COORDINATES

There are a few ways to express the velocity vector in rectangular coordinates. Below is a summary equation in two dimensions with interchangeable parts. Be sure to become comfortable with any of these versions. Take special note the convention of using "dot notation" which is often new to some students. The dot means "derivative with respect to time":

$$\vec{v} = \frac{d}{dt}\left(\vec{r}\right) = \frac{d}{dt}\left(x\hat{i} + y\hat{j}\right) = \left(\frac{dx}{dt}\right)\hat{i} + \left(\frac{dy}{dt}\right)\hat{j} = v_x\hat{i} + v_y\hat{j} = \dot{x}\hat{i} + \dot{y}\hat{j}.$$

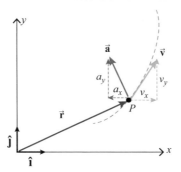

Figure 3.8: Rectangular coordinates components of velocity and acceleration.

Taken separately, the components are: $v_x = \frac{dx}{dt} = \dot{x}$ and $v_y = \frac{dy}{dt} = \dot{y}$.

The important concept: *in rectangular coordinates the velocity components are the time rate change of the components of the position vector.* As we'll see, this is not true for other coordinate systems.

Likewise, acceleration can be written in multiple ways:

$$\vec{\mathbf{a}} = \frac{d}{dt}\left(\vec{\mathbf{v}}\right) = \frac{d}{dt}\left(v_x\hat{\imath} + v_y\hat{\jmath}\right) = \left(\frac{dv_x}{dt}\right)\hat{\imath} + \left(\frac{dv_y}{dt}\right)\hat{\jmath} = a_x\hat{\imath} + a_y\hat{\jmath} = \ddot{x}\hat{\imath} + \ddot{y}\hat{\jmath}.$$

Figure 3.8 shows the components of the acceleration vector included with the position and velocity vectors.

Recall the other definitions of acceleration and there are even more ways to write out the components in rectangular coordinates:

$$a_x = \frac{dv_x}{dt} = v_x\frac{dv_x}{dx} = \frac{d^2x}{dt^2} = \ddot{x} \quad \text{and} \quad a_y = \frac{dv_y}{dt} = v_y\frac{dv_y}{dy} = \frac{d^2y}{dt^2} = \ddot{y}.$$

The important concept: *in rectangular coordinates the acceleration components are the time rate change of the components of the velocity vector.*

Note again, this is unique to rectangular coordinates. In other coordinate systems the acceleration components are NOT the time derivative of the velocity components.

Example 3.1

A particle travels according to the velocity $\vec{\mathbf{v}} = \left(45t^2 + 20t\right)\hat{\imath} + (40\cos 5t)\hat{\jmath}$ ft/s beginning at $\vec{\mathbf{r}}_0 = (2)\hat{\imath} + (1)\hat{\jmath}$ ft. Determine the time, position vector, velocity, and acceleration when the velocity is first completely horizontal? Plot the path and superimpose the velocity and acceleration vector components onto the graph.

The velocity is found by integration using $\vec{v} = \frac{d\vec{r}}{dt}$

$$\int_{r_0}^{r} d\vec{r} = \int_{0}^{t} \vec{v} \, dt = \int_{0}^{t} \left[(45t^2 + 20t)\,\hat{i} + (40\cos 5t)\,\hat{j} \right] dt$$

$$\vec{r} - \vec{r}_0 = (15t^3 + 10t^2)\,\hat{i} + (8\sin 5t)\,\hat{j}$$

$$\vec{r}(t) = (15t^3 + 10t^2 + 2)\,\hat{i} + (8\sin t + 1)\,\hat{j}.$$

The acceleration is found from differentiation:

$$\vec{a} = \frac{d\vec{v}}{dt} = \frac{d}{dt}\left[(45t^2 + 20t)\,\hat{i} + (40\cos 5t)\,\hat{j} \right]$$

$$\vec{a} = (90t + 20)\,\hat{i} + (-200\sin 5t)\,\hat{j} \text{ ft/s}^2.$$

The velocity is completely horizontal when $v_y = 0$, which occurs when $\cos 5t = 0$, so $t = \frac{\pi}{10}, \frac{3\pi}{10}, \frac{5\pi}{10}, \ldots$. The first time is when $t = \frac{\pi}{10} = 0.3142$ s.

The position at this time is:

$$\vec{r}(0.3142) = (3.452)\,\hat{i} + (9.000)\,\hat{j} \text{ ft}.$$

The velocity is:

$$\vec{v}(0.3142) = (10.72)\,\hat{i} + (0)\,\hat{j} \text{ ft/s}.$$

The acceleration is:

$$\vec{a}(0.3142) = (48.27)\,\hat{i} + (-200)\,\hat{j} \text{ ft/s}^2.$$

A plot of the path with velocity and acceleration vector components looks like Figure 3.9. Two key things to note: velocity is tangent to the path, and acceleration is not.

3.4   PROJECTILE MOTION

A special case of kinematics with rectangular coordinates is projectile motion. This is perhaps the "first kinematics problem" as our ancestors attempted to throw rocks at potential food and at one another. Most likely the best rock throwers were those who quickly recognized there was an arc to the path of the rock that required them to aim higher than the target. The arc, such as the one made by Wormy's apple in Figure 3.10, is caused by the difference in kinematics in the vertical and horizontal directions. This special case of kinematics has constant acceleration (gravity) in the vertical direction and constant velocity in two directions if we neglect aerodynamic losses (aerodynamic drag causes deceleration but is often small enough to be neglected in relatively slow speed projectiles).

In two dimensions the useful equations are:

$$v_x = \dot{x} = (v_x)_0 \qquad v_y = \dot{y} = (v_y)_0 - gt$$

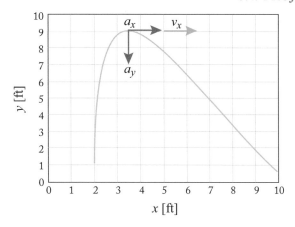

Figure 3.9: Example 3.1 results.

Figure 3.10: Newtdog pitches Wormy for projectile motion (©E. Diehl).

$$x = x_0 + (v_x)_0 t \qquad y = y_0 + (v_y)_0 t - \frac{1}{2}gt^2.$$

Note, it can *sometimes* also be useful to use $v_y{}^2 = (v_y)_0{}^2 - 2g(y - y_0)$, especially if the peak of the motion (when v_y is zero) is a useful aspect of the problem.

We can separate the motion into two simultaneous motions that are *almost* independent… their position is linked by time. We can even think of these as two separate motions happening simultaneously. In Figure 3.11 the arc of the path Wormy makes can be broken down into an up and down vertical motion (where the velocity changes) and a horizontal motion with constant velocity.

We can combine the projectile motion position equations to describe the path of the motion, resulting in a parabola. Solving the x position equation for time ($t = \frac{x - x_0}{(v_x)_0}$), substituting

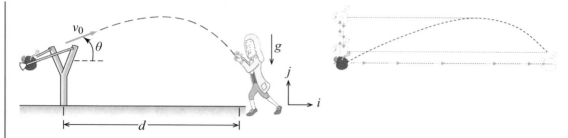

Figure 3.11: Projectile motion path is parabolic (©E. Diehl).

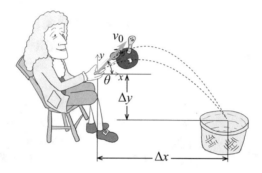

Figure 3.12: Newtdog tosses Wormy into basket for Example 3.2 (©E. Diehl).

into the y position equation, and treating the original position as $(x, y) = (0, 0)$ we obtain:

$$y = (v_y)_0 \frac{x}{(v_x)_0} - \frac{1}{2} g \left[\frac{x}{(v_x)_0} \right]^2 = \left[\frac{(v_y)_0}{(v_x)_0} \right] x - \left[\frac{g}{2(v_x)_0{}^2} \right] x^2.$$

The result is an equation with y as a function of x. A graph of y vs. x is the path the projectile takes.

Example 3.2

Newtdog wants to toss Wormy into the center of an apple basket positioned 2 ft below his release point and 6 ft away (Figure 3.12). Determine the required release speed in miles per hour if his release angle is 45° or 30°.

Define coordinate system at the release point $x_0 = 0$, $y_0 = 0$

The end point is therefore $x = 6$ ft $y = -2$ ft

Initial velocity components $(v_x)_0 = v_0 \cos \theta$ $(v_y)_0 = v_0 \sin \theta$

x-dir position relation at constant velocity $x = x_0 + (v_x)_0 t$

Solve for time $\quad t = \frac{x - x_0}{(v_x)_0}$ ①

y-dir position relation at constant acceleration $\quad y = y_0 + (v_y)_0 t - \frac{1}{2} g t^2$ ②

Plug ① into ② $\quad y = y_0 + \frac{(v_y)_0}{(v_x)_0}(x - x_0) - \frac{1}{2} g \frac{(x - x_0)^2}{(v_x)_0^2}$

Substitute the initial velocity components $\quad y = y_0 + \frac{v_0 \sin \theta}{v_0 \cos \theta}(x - x_0) - \frac{1}{2} g \frac{(x - x_0)^2}{v_0^2 \cos^2 \theta}$

Enter the knowns (keep the angle variable): $\quad y = (0) + \tan \theta \, (x - (0)) - \frac{1}{2}(32.2)\frac{(x - (0))^2}{v_0^2 \cos^2 \theta}$

Solve for initial velocity magnitude (speed): $\quad v_0 = \sqrt{\frac{\frac{1}{2}(32.2)x^2}{(x \tan \theta - y)\cos^2 \theta}}$

Find the initial speed at 45° $\quad v_0 = \sqrt{\frac{\frac{1}{2}(32.2)(6)^2}{((6)\tan(45°) - (-2))\cos^2(45°)}} = 12.04$ ft/s = 8.207 mph

Find the initial speed at 30° $\quad v_0 = \sqrt{\frac{\frac{1}{2}(32.2)(6)^2}{((6)\tan(30°) - (-2))\cos^2(30°)}} = 11.89$ ft/s = 8.109 mph

$$\boxed{v_0 = 8.21 \text{ mph or } 8.11 \text{ mph}}.$$

It's a good idea to convert answers into units you can relate to. In this case we can get a better sense of how fast these speeds are by converting them into miles per hour because we have several ways to reference how much that is, for example a major league pitcher who is very good can throw a 100 mph fast ball or the record 40-yard dash speed is about 19.4 mph.

It's also useful to plot the path of each result to see how two different angle and speed combinations will achieve the same results as in Figure 3.13. It's critical to keep track of what's actually taking place in all dynamics problems. Visualizing is a key part of this.

While projectile motion problems may seem simple based on the provided examples, variations of knowns, unknowns, and arrangements can make them quite challenging because it requires us to think through the scenarios rather than memorize all of the permutations of this fundamental problem. Consider the things you might want to use projectile motion to accomplish or know about. Below is a list of some possibilities.

- The distance a projectile will travel.

- The peak height a projectile will go.

- The angle(s) required at launch to land at a particular spot (there are often two).

- The velocity required to land at a particular spot.

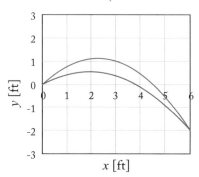

Figure 3.13: Graph of path in Example 3.2.

- The time at which a projectile will reach a particular spot.

- The location (or time) when two projectiles collide.

- The velocity when a projectile lands (including angle of approach).

- The distance a projectile will go if it intersects with a slope.

This is not a complete list; there are other possible projectile motion problems. Considering the potential complexity of a projectile motion problem, it is remarkable when athletes train themselves to make split second adjustments to hit a target with a sports ball, as in Newtdog's drop kick in Figure 3.14. The important take away from the above list: you can't simply memorize a step-by-step solution method in Dynamics, you will have to think through problems and figure out a solution strategy.

3.5   RELATIVE MOTION WITH VECTORS

In Section 2.2 we discussed one-dimensional relative motion using the convention of sub-scripts which are "with respect to." We can extend this to two and three dimensions (although we'll restrict our discussion to two-dimensions for convenience). We begin with the position vector and what we know about vector addition to describe relative position. The location of point B (with respect to the origin, O) can be determined using a known position of point A (with respect to the origin) plus the relative position of B with respect to A, written as: $\vec{r}_{B/O} = \vec{r}_{A/O} + \vec{r}_{B/A}$ and shown in Figure 3.15. This can also be rearranged to check the order of relation: $\vec{r}_{B/A} = \vec{r}_{B/O} - \vec{r}_{A/O}$. Note there is a sort of indices math here, where the O factors out. Also note we could alternatively use point B as the reference to determine point A, in which case the sign (and direction) of the relative position vector reverses as does the order of the letters in the index (i.e., $\vec{r}_{A/B} = -\vec{r}_{B/A}$).

Figure 3.14: Newtdog drop kicks a football centuries before his time (©E. Diehl).

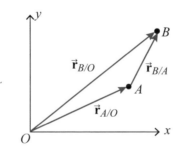

Figure 3.15: Relative position vector.

The origin is often a fixed point, so we leave off this reference and write the position vector of point B as:

$$\vec{r}_B = \vec{r}_A + \vec{r}_{B/A}.$$

The derivative of this position vector leads to the velocity vector:

$$\vec{v}_B = \vec{v}_A + \vec{v}_{B/A}.$$

And taking the derivative of the velocity vector:

$$\vec{a}_B = \vec{a}_A + \vec{a}_{B/A}.$$

In all of these cases, point A is used as a reference to determine point B. The vector addition discussed in Section 3.1 is especially useful here. We will use this concept extensively for rigid body kinematics.

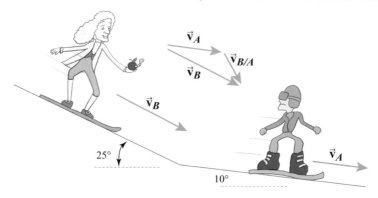

Figure 3.16: Relative motion in Example 3.3 (©E. Diehl).

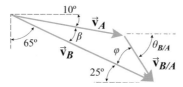

Figure 3.17: Triangle for Example 3.3.

Example 3.3

Two snow boarders are traveling down a hill that changes slope from 25°–10° (Figure 3.16). Newdog (snowboarder B) is traveling at 14 m/s on the steeper slope while his new friend is going 10 m/s. Determine the relative velocity of Newtdog with respect to the other skier.

There are always two ways to approach vector relative motion problems: coordinates or triangles:

$$\vec{v}_A = 10 \text{ m/s} \searrow 10° = \left((10)\cos\left(10°\right)\right)\hat{i} + \left(-(10)\sin\left(10°\right)\right)\hat{j} = (9.848)\hat{i} + (-1.736)\hat{j}$$
$$\vec{v}_B = 14 \text{ m/s} \searrow 25° = \left((14)\cos\left(25°\right)\right)\hat{i} + \left(-(14)\sin\left(25°\right)\right)\hat{j} = (12.69)\hat{i} + (-5.917)\hat{j}$$
$$\vec{v}_{B/A} = \vec{v}_B - \vec{v}_A = [(12.69)\hat{i} + (-5.917)\hat{j}] - [(9.848)\hat{i} + (-1.736)\hat{j}]$$
$$= (2.840)\hat{i} + (-4.181)\hat{j}$$
$$\left|\vec{v}_{B/A}\right| = \sqrt{(2.840)^2 + (-4.181)^2} = 5.054 \text{ m/s} \quad \theta_{B/A} = \tan^{-1}\left(\frac{-4.181}{2.840}\right) = -55.81°$$

$$\boxed{\vec{v}_{B/A} = 5.05 \text{ m/s} \searrow 55.8°} .$$

Another approach is to use vector triangles (Figure 3.17). The geometry can be frustrating, but this is a very good way to visualize the results. By inspection the interior angle is $\beta = 15°$.

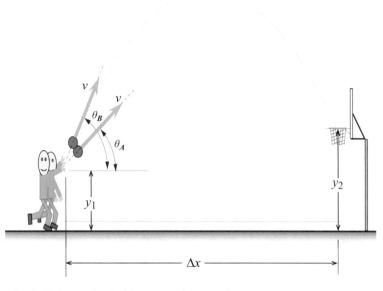

Figure 3.18: Basketball from the half court in Example 3.4.

Law of Cosines to get magnitude of $\left|\vec{v}_{B/A}\right|$:

$$c^2 = a^2 + b^2 - 2ab \cos C$$

$$\left|\vec{v}_{B/A}\right| = \sqrt{\left|\vec{v}_A\right|^2 + \left|\vec{v}_B\right|^2 - 2\left|\vec{v}_A\right|\left|\vec{v}_B\right| \cos \beta}$$

$$\left|\vec{v}_{B/A}\right| = \sqrt{(10)^2 + (14)^2 - 2\,(10)\,(14) \cos\,(15°)} = 5.054 \text{ m/s}.$$

Law of Sines to get angle: $\frac{a}{\sin A} = \frac{c}{\sin C}$

$$\frac{(10)}{\sin \varphi} = \frac{(5.054)}{\sin\,(15°)} \qquad\qquad \varphi = 30.80°.$$

The angle of the vector is $\theta = 25° + 30.80° = 55.80°$ which confirms the vector addition.

Example 3.4
Twin brothers shoot basketballs simultaneously from the half court, as shown in Figure 3.18. Both make the basket launching with the same initial speed but brother B launches his ball from a steeper angle than brother A, and therefore his ball goes into the basket after brother A's ball. Half court to the center of the basket in a professional basketball court is $\Delta x = 41.75$ ft, and the rim is $y_2 = 10$ ft high. The brothers launch the ball from $y_1 = 6$ ft above the floor at

a speed of $v_1 = 40$ ft/s. What is the relative velocity of ball B to ball A when ball A enters the hoop.

We start by setting up equations for just one basketball.

x-dir position relation at constant velocity $x_2 = x_1 + v_1 \cos \theta t$

Solve for time $t = \frac{x_2 - x_1}{v_1 \cos \theta}$ ①

y-dir position relation at constant acceleration $y_2 = y_1 + v_1 \sin \theta t - \frac{1}{2} g t^2$ ②

Plug ① into ② $y_2 = y_1 + \frac{v_1 \sin \theta}{v_1 \cos \theta} (x_2 - x_1) - \frac{1}{2} g \frac{(x_2 - x_1)^2}{v_1{}^2 \cos^2 \theta}$

$$y_2 = y_1 + \tan \theta (x_2 - x_1) - \frac{1}{2} g \frac{(x_2 - x_1)^2}{v_1{}^2 \cos^2 \theta}.$$

A bit of digging into trigonometry identities uncovers that $\frac{1}{\cos^2 \theta} = \sec^2 \theta = 1 + \tan^2 \theta$

$$y_2 = y_1 + \tan \theta (x_2 - x_1) - \left[\frac{1}{2} g \frac{(x_2 - x_1)^2}{v_1{}^2} \right] (1 + \tan^2 \theta).$$

Substitute the knowns $v_1 = 40$ ft/s, $x_1 = 0$, $x_2 = 41.75$ ft, $y_1 = 6$ ft, and $y_2 = 10$ ft.

$$(10) = (6) + \tan \theta (41.25) - \left[\frac{1}{2} (32.2) \frac{(41.25)^2}{(40)^2} \right] (1 + \tan^2 \theta)$$

$$(4) = (41.25) \tan \theta - (17.12) - (17.12) \tan^2 \theta$$

$$(17.12) \tan^2 \theta - (41.25) \tan \theta + (21.12) = 0.$$

We temporarily replace $z = \tan \theta$ and find the roots:

$$(17.12) z^2 - (41.25) z + (21.12) = 0$$

$$z = \frac{-b \pm \sqrt{b^2 - 4ac}}{2a} = \frac{-(-41.25) \pm \sqrt{(-41.25)^2 - 4 (17.12) (21.12)}}{2 (17.12)} = 1.671, \ 0.7381$$

$$\theta_A = \tan^{-1} (0.7381) = 36.43°, \quad \theta_B = \tan^{-1} (1.671) = 59.11°.$$

These are the two launch angles. Ball A reaches the basket at:

$$t_{A2} = \frac{x_2 - x_1}{v_1 \cos \theta_A} = \frac{(41.25)}{(40) \cos (36.43°)} = 1.282 \text{ s}.$$

At this time the velocity components of ball A are:

$$(v_A)_x = v_1 \cos \theta_A = (40) \cos (36.43°) = 32.18 \text{ ft/s}$$

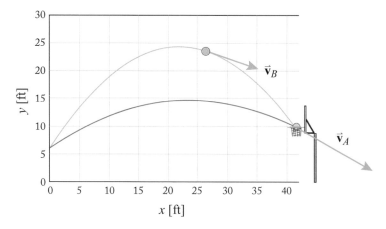

Figure 3.19: Basketball paths of Example 3.4.

$$(v_{A2})_y = (v_{A1})_y - gt_2 = (40)\sin\left(36.43°\right) - (32.2)(1.282\ ) = -17.53 \text{ ft/s}.$$

The velocity components of ball B at this time are

$$(v_B)_x = v_1 \cos\theta_B = (40)\cos\left(59.11°\right) = 20.54 \text{ ft/s}$$

$$(v_{B2})_y = (v_{B1})_y - gt_2 = (40)\sin\left(59.11°\right) - (32.2)(1.282) = -6.954 \text{ ft/s}.$$

The relative velocity is found from:

$$\vec{\mathbf{v}}_A = (32.18)\,\hat{\imath} + (-17.53)\,\hat{\jmath}$$

$$\vec{\mathbf{v}}_B = (20.54)\,\hat{\imath} + (-6.954)\,\hat{\jmath}$$

$$\vec{\mathbf{v}}_{B/A} = \vec{\mathbf{v}}_B - \vec{\mathbf{v}}_A = [(32.18)\,\hat{\imath} + (-17.53)\,\hat{\jmath}] - [(20.54)\,\hat{\imath} + (-6.954)\,\hat{\jmath}]$$
$$= (11.64)\,\hat{\imath} + (-10.58)\,\hat{\jmath}$$

$$\left|\vec{\mathbf{v}}_{B/A}\right| = \sqrt{(11.64)^2 + (-10.58)^2} = 15.73 \text{ ft/s} \quad \theta_{B/A} = \tan^{-1}\left(\frac{-10.58}{11.64}\right) = -42.27°$$

$$\boxed{\vec{\mathbf{v}}_{B/A} = 15.7 \text{ ft/s} \searrow 42.3°}\ .$$

Figure 3.19 shows the paths of each ball and the location of the balls when ball A enters the hoop. Graphing the trajectory of both balls, the location of ball B when ball A goes through the hoop, and the tangents to represent the velocity directions confirms the numeric calculations.

Book 1 - Class 4

https://www.youtube.com/watch?v=vr5ZudtUhD0

CLASS 4

Non-Rectangular Coordinate Systems: Path Coordinates

B.L.U.F. (Bottom Line Up Front)

- Coordinate Transformation: project existing unit vectors onto new unit vectors by using sines and cosines in a matrix.

- Path Coordinates (a.k.a. "Tangential and Normal"): attached to the particle.

- Velocity is expressed in Path Coordinates as: $\vec{\mathbf{v}} = v\hat{\mathbf{e}}_t$.

- Acceleration is expressed in Path Coordinates as: $\vec{\mathbf{a}} \left(\frac{dv}{dt}\right)\hat{\mathbf{e}}_t + \left(\frac{v^2}{\rho}\right)\hat{\mathbf{e}}_n$.

4.1   WHY USE NON-RECTANGULAR COORDINATE SYSTEMS?

Most of us are familiar with the rectangular coordinate system (a.k.a. "Cartesian Coordinates") and it may seem like an unnecessary hassle to use anything else. Hopefully, you will soon see there are advantages to using other types of coordinate systems. First, let's remember that we generally use "orthogonal" coordinate systems whose unit vectors are perpendicular to one another, and the three unit vectors follow the "right-hand rule" or orientation. The two non-rectangular coordinate systems we'll use are Path (Figure 4.1) and Polar (Figure 4.2). These images will be repeated and their significance explained in subsequent sections.

There can be other coordinate systems, but those are for peculiar situations… which begs the question: "why?" The simple answer is that the geometry of certain motions "fit" more easily in certain coordinates. An example might be the motion of satellites orbiting the Earth. Spherical coordinates are often used since the motion of the satellite is like traveling on the surface of a sphere.

In more complicated physics courses, such as Continuum Mechanics, problems are solved with generalized coordinates that disregard geometry so the math is somewhat simplified in that some algebra and trigonometry is avoided. Similarly, we use non-rectangular coordinates in this course in Dynamics to avoid (some of) the algebra and trigonometry necessary in rectangular

Figure 4.1: Newtdog stays tangent to the path in path coordinates (©E. Diehl).

Figure 4.2: Newtdog takes Wormy for a spin in polar coordinates (©E. Diehl).

coordinates. It also turns out that we can often gain insight into the motion by using these other coordinate systems.

4.2 COORDINATE SYSTEM TRANSFORMATION

We often want to switch between coordinate systems or create a localized reference coordinate system. The most systematic approach is to apply a transformation matrix. We'll use 2-D for these examples where one can also overlay the coordinates and work through the sines and cosines, but be aware that for three dimensions matrix math is required for coordinate transformation. Note that coordinate transformation is a very necessary part of the math behind video games and computer-generated imagery, so understanding this simplified procedure has more advanced applications that you might encounter elsewhere.

Figure 4.3: Setting up a unit vector transformation.

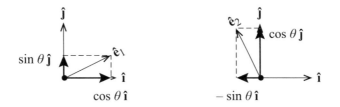

Figure 4.4: Projecting each new unit vector onto old.

Assume we begin with rectangular coordinates with unit vectors $\hat{\imath}$ and $\hat{\jmath}$ and want to switch to a new set of orthogonal unit vectors $\hat{e}_1$ and $\hat{e}_2$, which make up the new generic coordinate system. We over-lay the two sets and find the angle formed between them, as in Figure 4.3. Remember that while unit vectors are often associated with a point (such as the origin or the moving particle as we'll see in the two coordinate systems discussed in this class), they only define directions and are not actually fixed onto points, so they can be moved around and overlaid as shown.

Take each new unit vector separately and project the new unit vector onto the old ones as in the equations below corresponding to Figure 4.4. We see there is a portion in each of the old coordinate system. The equations for each new vector are:

$$\hat{e}_1 = \cos\theta\hat{\imath} + \sin\theta\hat{\jmath}$$
$$\hat{e}_2 = -\sin\theta\hat{\imath} + \cos\theta\hat{\jmath}.$$

This is a step-by-step explanation, and it isn't necessary to draw each of the vectors this way each time, but it's good to think of them this way. We can rearrange these equations into a matrix equation:

$$\left\{ \begin{array}{c} \hat{e}_1 \\ \hat{e}_2 \end{array} \right\} = \left[\begin{array}{cc} \cos\theta & \sin\theta \\ -\sin\theta & \cos\theta \end{array} \right] \left\{ \begin{array}{c} \hat{\imath} \\ \hat{\jmath} \end{array} \right\}.$$

The 2×2 matrix is the "transformation matrix" and has a few noteworthy traits. It is *almost* symmetrical diagonally except one of the values is the opposite sign of the others. There are several possible variations depending on the situation, so it's best to think through the formulation of the matrix rather than memorize each version.

Table 4.1: Coordinate transformation array example

	$\hat{\imath}$	$\hat{\jmath}$
$\vec{e}_1$	$\cos\theta$	$\sin\theta$
$\vec{e}_2$	$-\sin\theta$	$\cos\theta$

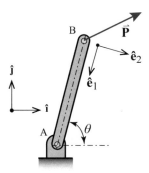

Figure 4.5: Example 4.1 linkage.

It is often convenient to follow a procedure whereby you construct a table called a "coordinate transformation array," as in Table 4.1, which can be read either by columns to transform into the row coordinate system, or by rows to transform into the column coordinate system.

The following example will help demonstrate the procedure and show that the transformation isn't always identical to the above, especially if one avoids using angles greater than 90° to define the rotation.

Example 4.1
Vector $\vec{P} = (4)\hat{\imath} + (2)\hat{\jmath}$ is associated with linkage AB shown in Figure 4.5. Find the coordinate transformation array from $\hat{\imath}$ and $\hat{\jmath}$ to $\vec{e}_1$ and $\vec{e}_2$ and express vector $\vec{P}$ in this new coordinate system when $\theta = 75°$.

We find the new coordinate system unit vectors, $\vec{e}_1$ and $\vec{e}_2$, in terms of the original unit vectors ($\hat{\imath}$ and $\hat{\jmath}$) and then arrange these into the transformation array in Table 4.2:

$$\hat{e}_1 = -\cos\theta\hat{\imath} - \sin\theta\hat{\jmath}$$

$$\hat{e}_2 = \sin\theta\hat{\imath} - \cos\theta\hat{\jmath}.$$

To use this array, we multiply the $\hat{\imath}$ component downward by the first column and the $\hat{\jmath}$ component by the second column:

$$\vec{P} = (4)(-\cos\theta\hat{e}_1 + \sin\theta\hat{e}_2) + (2)(-\sin\theta\hat{e}_1 - \cos\theta\hat{e}_2)$$

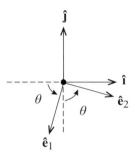

Figure 4.6: Example 4.1 coordinate axes.

Table 4.2: Example 4.1 Coordinate transformation array

	$\hat{\mathbf{i}}$	$\hat{\mathbf{j}}$
$\vec{\mathbf{e}}_1$	$-\cos\theta$	$-\sin\theta$
$\vec{\mathbf{e}}_2$	$\sin\theta$	$-\cos\theta$

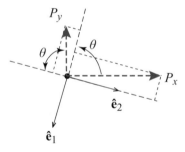

Figure 4.7: Example 4.1 force components.

$$\overrightarrow{\mathbf{P}} = \left[-\left(4\right)\cos\left(75°\right) - \left(2\right)\sin\left(75°\right)\right]\hat{\mathbf{e}}_1 + \left[\left(4\right)\sin\left(75°\right) - \left(2\right)\cos\left(75°\right)\right]\hat{\mathbf{e}}_2$$

$$\boxed{\overrightarrow{\mathbf{P}} = \left(-2.967\right)\hat{\mathbf{e}}_1 + \left(3.346\right)\hat{\mathbf{e}}_2}\,.$$

One could have also broken this vector into its components and projected them onto the new coordinate system as depicted in Figure 4.7:

$$\overrightarrow{\mathbf{P}} = \left[-P_x\cos\theta - P_y\sin\theta\right]\hat{\mathbf{e}}_1 + \left[P_x\sin\theta - P_y\cos\theta\right]\hat{\mathbf{e}}_2$$

$$\overrightarrow{\mathbf{P}} = \left[-\left(4\right)\cos\left(75°\right) - \left(2\right)\sin\left(75°\right)\right]\hat{\mathbf{e}}_1 + \left[\left(4\right)\sin\left(75°\right) - \left(2\right)\cos\left(75°\right)\right]\hat{\mathbf{e}}_2$$

$$\boxed{\overrightarrow{\mathbf{P}} = \left(-2.967\right)\hat{\mathbf{e}}_1 + \left(3.346\right)\hat{\mathbf{e}}_2}\,.$$

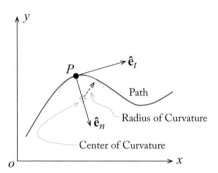

Figure 4.8: **Path coordinates unit vectors.**

Either method can be used, so when time permits it's advisable to double check your results by using both since it is easy to get signs reversed and sines and cosines mixed. Again, coordinate transformation is often done within computer codes since there can be many frames of reference to consider.

4.3   PATH COORDINATES (A.K.A. "TANGENTIAL AND NORMAL")

When trying to describe motion of a particle, we sometimes want to do it as if we are an observer from afar while other times it's more convenient to travel along with the particle. In the latter case, we pretend we're traveling with the particle along its path and assign tangential and normal unit vectors ($\hat{\mathbf{e}}_t$ and $\hat{\mathbf{e}}_n$) associated with that path at a particular time, as shown in Figure 4.8. An important feature of path coordinates to note is the unit vectors are always rotating if the path is not a straight line and the normal points inward toward the center of curvature remaining orthogonal to the tangential.

This is a *local coordinate system* (as opposed to a a fixed global coordinate system) so it changes with the particle. Also, it has no position vector because there is no fixed point to reference.

Figure 4.9 shows the velocity and acceleration vectors overlaid on the point. The velocity vector of the particle in path coordinates is written as:

$$\boxed{\vec{\mathbf{v}} = v\hat{\mathbf{e}}_t}\,.$$

As you'll recall, velocity is ALWAYS tangent to the path, so the magnitude of the velocity (a.k.a. "speed") v, times the tangential unit vector, is the velocity. You should also recall that acceleration is NOT ALWAYS tangent to the path. When is it tangent to the path? When the path is a straight line. Otherwise there is always a normal component of acceleration.

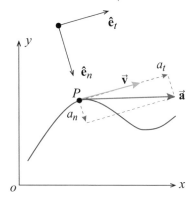

Figure 4.9: Path coordinates velocity and acceleration components.

The acceleration vector in path coordinates is written as:

$$\vec{\mathbf{a}} = \left(\frac{dv}{dt}\right)\hat{\mathbf{e}}_{t} + \left(\frac{v^2}{\rho}\right)\hat{\mathbf{e}}_n .$$

It can also be written as $\boxed{\vec{\mathbf{a}} = a_t\hat{\mathbf{e}}_t + a_n\hat{\mathbf{e}}_n}$,
where:

v = speed

ρ = radius of curvature

a_t = tangential acceleration = change in speed = $\frac{dv}{dt} = \dot{v}$

a_n = normal acceleration = "centrifugal" or "centripetal" due to change in direction
= $\frac{v^2}{\rho}$

Why does the acceleration have an additional unit vector when velocity only has tangential?

There are a couple of ways to explain normal acceleration. Here is a scenario that can help us to understand it. The cartoon in Figure 4.10 of Newtdog cycling off a cliff is meant to serve as a mnemonic device to help you remember the additional component, normal acceleration (a.k.a. "centrifugal acceleration") and why it's necessary.

This is a road that curves to the left. What is Newtdog's speed to the left before he enters the curve? Zero. If he wants to stay on the curve, he needs some speed in the left direction. Therefore his speed to the left must increase from zero to something. That is his normal acceleration.

Without normal acceleration, he would continue off the curve and off the cliff. Try to picture this image when dealing with path coordinates.

Figure 4.10: Newtdog demonstrating the importance of having normal acceleration (repeat of Figure 4.1) (©E. Diehl).

A note on "centrifugal" vs. "centripetal"… this is a semantic distinction made when discussing apparent vs. reaction forces due to this acceleration. Call it whatever you'd like as long as you recognize what it is.

Let's see mathematically why we get an extra unit vector with acceleration compared to velocity. The velocity vector is $\vec{\mathbf{v}} = v\hat{\mathbf{e}}_t$, and acceleration is the time derivative of this. The product rule of differentiation requires that both the magnitude (speed) and the unit vector are differentiated if they change with time:

$$\frac{d\vec{\mathbf{v}}}{dt} = \frac{d\,(v\hat{\mathbf{e}}_t)}{dt} = \frac{dv}{dt}\hat{\mathbf{e}}_t + v\frac{d\,(\hat{\mathbf{e}}_t)}{dt}.$$

The question arises, what is $\frac{d(\hat{\mathbf{e}}_t)}{dt}$? The unit vector does change with time, so is not zero. To figure this out we consider the motion of two points along the path, P and P'. Figure 4.11 shows the tangential unit vector for each point, $\hat{\mathbf{e}}_t$ and $\hat{\mathbf{e}}_t'$.

The change in the unit vectors after the point moves from P to P' is $\Delta\hat{\mathbf{e}}_t$. A vector triangle representation of this change is shown within the figure. We begin to see that the change of the unit vector points inward of the curvature of the path.

Lines drawn perpendicular to the unit vector merge at a point with an angular change between them of $\Delta\theta$. The angular rate of change with the unit vector change, as it gets smaller and smaller points the unit vector toward the center of the curvature, which we know is the

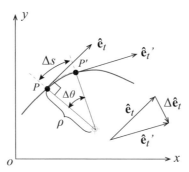

Figure 4.11: Consequences of moving unit vectors in path coordinates.

normal unit vector, expressed mathematically as:

$$\lim_{\Delta\theta \to 0} \frac{\Delta \hat{\mathbf{e}}_t}{\Delta\theta} = \frac{d\,(\hat{\mathbf{e}}_t)}{d\theta} = \hat{\mathbf{e}}_n.$$

To determine the relationship between $\frac{d(\hat{\mathbf{e}}_t)}{d\theta}$ and $\frac{d(\hat{\mathbf{e}}_t)}{dt}$ we apply the chain rule: $\frac{d(\hat{\mathbf{e}}_t)}{dt} = \frac{d(\hat{\mathbf{e}}_t)}{d\theta}\frac{d\theta}{ds}\frac{ds}{dt}$.

From Figure 4.11 we note that the length of the arc swept by the change along the path is $\Delta s = \rho\Delta\theta$, which when smaller can be as written $ds = \rho\,d\theta$ or rearranged to $\frac{ds}{d\theta} = \frac{1}{\rho}$. We also observe that the change in arc length with time is speed, so $\frac{ds}{dt} = v$. Since we established above that the change in the tangential unit vector with angle is $\frac{d(\hat{\mathbf{e}}_t)}{d\theta} = \hat{\mathbf{e}}_n$, the conclusion is: $\frac{d(\hat{\mathbf{e}}_t)}{dt} = \hat{\mathbf{e}}_n\frac{1}{\rho}v$. When applied to the time derivative of the velocity vector (acceleration), we get the previously stated equation for acceleration in path coordinates:

$$\boxed{\vec{\mathbf{a}} = \left(\frac{dv}{dt}\right)\hat{\mathbf{e}}_t + \left(\frac{v^2}{\rho}\right)\hat{\mathbf{e}}_n}.$$

The above could be considered a "proof," but those are generally more formal in mathematical lingo. Consider this an explanation that should help you understand and remember these very important points about acceleration in path coordinates.

There are two parts to the acceleration:

1. the change in speed acting along the tangent of the path: $\frac{dv}{dt}$; and

2. the change in direction acting toward the center of curvature of the path: $\frac{v^2}{\rho}$.

We will discuss motion described in the polar coordinate system in Class 5 using the same explanation process. Then we'll compare these two coordinate systems to see that the same motion can be described using either.

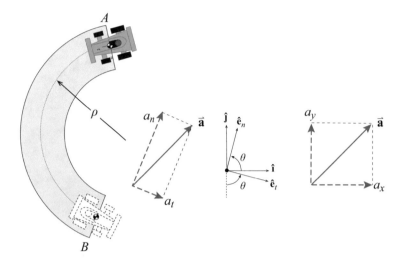

Figure 4.12: Example 4.2 race car.

Example 4.2

A race car travels from point A to point B along a curved section of raceway with 100-yard radius (Figure 4.12). The two points are 150° apart if the curve is approximately an arc of a circle. The race car speed is measured to be 100 mph at point A and 250 mph at point B. Determine the acceleration in ft/s² at point B in path and rectangular coordinates (looking down onto the race track and assuming the arc is symmetric as shown). Assume the rate of speed change is constant throughout.

First, let's convert the speeds into ft/s:

$$v_A = \frac{(100 \text{ mph}) (5,280 \text{ ft/mi})}{(3600 \text{ s/hr})} = 146.7 \text{ ft/s.}$$

$$v_B = \frac{(250 \text{ mph}) (5,280 \text{ ft/mi})}{(3600 \text{ s/hr})} = 366.7 \text{ ft/s.}$$

The distance traveled is $\Delta s = \rho \, \Delta\theta = (300 \text{ ft}) \left[\frac{(150°)(2\pi)}{(360°)}\right] = 785.4$ ft.

The last sentence of the question should be read carefully. It does not say "constant acceleration" which would apply to the entire acceleration, but instead is describing the tangential component of acceleration as constant. So, we can use the constant acceleration kinematic relationship:

$$v^2 = v_0{}^2 + 2a\Delta s$$

$$a_t = \frac{v^2 - v_0{}^2}{2\Delta s} = \frac{(366.7)^2 - (146.7)^2}{2(785.4)} = 71.89 \text{ ft/s}^2.$$

Table 4.3: Coordinate transformation array for Example 4.2

	$\hat{\imath}$	$\hat{\jmath}$
$\vec{e}_t$	$\sin\theta$	$-\cos\theta$
$\vec{e}_n$	$\cos\theta$	$\sin\theta$

The normal acceleration (due to change in direction) at point B is:

$$a_n = \frac{v^2}{\rho} = \frac{(366.7)^2}{(300)} = 448.2 \text{ ft/s}^2.$$

So the acceleration in path coordinates is written:

$$\boxed{\vec{a} = (71.9)\,\hat{e}_t + (448)\,\hat{e}_n \text{ ft/s}^2}.$$

To convert into rectangular components, we create a coordinate transformation array with $\theta = 75°$ (since we're told the section of 150° arc is symmetric). See the Cartesian and path coordinate axes in Figure 4.12 use to develop the transformation array. Because we are converting from Path to Cartesian we'll use by row rather than by column (Table 4.3):

$$\vec{a} = (71.89)\,[\sin\theta\hat{\imath} - \cos\theta\hat{\jmath}] + (448.2)\,[\cos\theta\hat{\imath} + \sin\theta\hat{\jmath}]$$

$$\vec{a} = \left[(71.89)\sin(75°) + (448.2)\cos(75°)\right]\hat{\imath} + \left[-(71.89)\cos(75°) + (448.2)\sin(75°)\right]\hat{\jmath}$$

$$\boxed{\vec{a} = (185.4)\,\hat{\imath} + (414.4)\,\hat{\jmath} \text{ ft/s}^2}.$$

We'll need the velocity at B in Cartesian coordinates for Example 5.2. In Path coordinates the velocity is $\vec{v} = (366.7)\,\hat{e}_t$ ft/s. Using the coordinate array we find it to be:

$$\vec{v} = (366.7)\left[\sin(75°)\hat{\imath} - \cos(75°)\hat{\jmath}\right] = (354.2)\hat{\imath} - (94.91)\hat{\jmath} \text{ ft/s}.$$

Last, it's often good to check the magnitude and direction of the acceleration to see if they make sense, especially the direction compared to the sketch:

$$|\vec{a}| = \sqrt{(185.4)^2 + (414.4)^2} = 454.0 \text{ ft/s}^2 \quad \theta_{\vec{a}} = \tan^{-1}\frac{(414.4)}{(185.4)} = 65.90°.$$

So the acceleration can also be written as: $\vec{a} = 454 \text{ ft/s}^2 \nearrow 65.9°$, which appears to be the expected direction. Note that the magnitude of acceleration is over 14 times the acceleration of gravity (g's) which is ridiculously large, although driving 250 mph around a curve isn't typically done even in race cars.

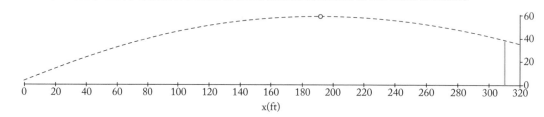

Figure 4.13: Baseball trajectory in Example 4.3.

Last, note that this form of the answer is sometimes called "polar notation" but this is NOT the same as using the Polar coordinate system, as we'll see in the next Class.

Example 4.3   A baseball is hit over a famous very tall outfield wall, barely making it over as shown in Figure 4.13. The wall is 37 ft, 2 in tall and 310 ft from home plate. The ball leaves the face of the bat 3 ft off the ground at a 30° angle from horizontal. Determine the radius of curvature of the ball's path at its peak and as it clears the wall.

When we recognize this is a projectile motion problem, we note that we're not given the initial speed of the ball, so that is our first goal similar to Example 3.2.

Starting point: $x_0 = 0$ ft $y_0 = 3$ ft.

Finishing point: $x_1 = 310$ ft $y_1 = 37'2'' = 37.17$ ft.

Initial velocity components: $(v_x)_0 = v_0 \cos \theta$ $(v_y)_0 = v_0 \sin \theta$.

x-dir position relation at constant velocity: $x = x_0 + (v_x)_0 t$.

Solve for time: $t = \frac{x - x_0}{(v_x)_0}$ ①.

y-dir position relation at constant acceleration: $y = y_0 + (v_y)_0 t - \frac{1}{2} g t^2$ ②.

Plug ① into ②: $y = y_0 + \frac{(v_y)_0}{(v_x)_0}(x - x_0) - \frac{1}{2} g \frac{(x - x_0)^2}{(v_x)_0^2}$.

Substitute the initial velocity component: $y = y_0 + \frac{v_0 \sin \theta}{v_0 \cos \theta}(x - x_0) - \frac{1}{2} g \frac{(x - x_0)^2}{v_0^2 \cos^2 \theta}$.

Solve for initial velocity magnitude (speed):

$$v_0 = \sqrt{\frac{\frac{1}{2}(32.2)(x - x_0)^2}{((x - x_0)\tan\theta + y_0 - y)\cos^2\theta}}$$

$$= \sqrt{\frac{\frac{1}{2}(32.2)(310)^2}{((310)\tan(30°) + (3) - (37.17))\cos^2(30°)}} = 119.4 \text{ ft/s} = 81.38 \text{ mph}.$$

The initial velocity components:

$$(v_x)_0 = v_0 \cos \theta = (119.4) \cos(30°) = 103.4 \text{ ft/s} \quad \rightarrow$$
$$(v_y)_0 = v_0 \sin \theta = (119.4) \sin(30°) = 59.68 \text{ ft/s} \quad \uparrow.$$

The velocity in the x-direction is constant. When the ball is at its peak its velocity is entirely horizontal. The only acceleration in projectile motion, neglecting aerodynamic drag, is from gravity. Since at the peak this acceleration is perpendicular to the velocity (which is tangential to the path), the normal acceleration equals gravity. Using this we can find the radius of curvature:

$$a_n = \frac{v^2}{\rho} = g$$

$$\rho = \frac{v^2}{g} = \frac{(103.4)^2}{(32.2)} = 331.8 \text{ ft}.$$

The velocity in the y-direction when the ball goes over the wall is:

$$(v_y)_1^2 = (v_y)_0^2 - 2g(y_1 - y_0)$$
$$(v_y)_1 = \sqrt{(v_y)_0^2 - 2g(y_1 - y_0)}$$
$$= \sqrt{(103.4)^2 - 2(32.2)((37.17) - (3))} = 36.89 \text{ ft/s} \downarrow.$$

The velocity at that moment is:

$$\vec{v}_1 = (103.4)\hat{\imath} + (-36.89)\hat{\jmath} \text{ ft/s}.$$

The magnitude and angle of the velocity are:

$$v_1 = \sqrt{(103.4)^2 + (-36.89)^2} = 109.8 \text{ ft/s} \quad \theta_1 = \tan^{-1}\frac{(36.89)}{(103.4)} = 19.64° \searrow.$$

To find the normal acceleration we need to break the acceleration of gravity into normal and tangential components. Even though we could do this in a single step, we'll get practice using the coordinate transformation we introduced in this Class. Figure 4.14 shows the overlaid unit vectors at the moment the ball goes over the wall. Table 4.4 is the corresponding transformation matrix.

The acceleration of gravity expressed as a vector is: $\vec{a} = (0)\hat{\imath} + (-32.2)\hat{\jmath} \text{ ft/s}^2$.

Applying the transformation matrix to this yields:

$$\vec{a} = (0)[\cos \theta \hat{e}_t - \sin \theta \hat{e}_n] + (-32.2)[-\sin \theta \hat{e}_t - \cos \theta \hat{e}_n]$$

$$\vec{a} = -(-32.2) \sin(19.64)\hat{e}_t - (-32.2) \cos(19.64°)\hat{e}_n$$

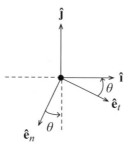

Figure 4.14: Unit vectors of Example 4.3.

Table 4.4: Coordinate transformation array of Example 4.3

	$\hat{\mathbf{i}}$	$\hat{\mathbf{j}}$
$\vec{\mathbf{e}}_t$	$\cos\theta$	$-\sin\theta$
$\vec{\mathbf{e}}_n$	$-\sin\theta$	$-\cos\theta$

$$\vec{a} = (10.82)\hat{e}_t + (30.33)\hat{e}_n \text{ ft/s}^2.$$

To find the radius of curvature we use the normal component:

$$a_n = \frac{v^2}{\rho} \qquad \rho = \frac{v^2}{g} = \frac{(109.8)^2}{(30.33)} = 397.5 \text{ ft.}$$

It makes sense that the radius at the wall would be larger than the at the peak.
The answers are: $\rho_{peak} = 332$ ft and $\rho_{wall} = 398$ ft.

Look back over this problem and reflect on the thought process required to formulate a solution strategy and apply the things we know.

Example 4.4   The rocket in Figure 4.15 is traveling horizontally at 100 m/s when the first stage falls away and the second stage ignites, propelling it along a path that follows $y = 0.1x^2$ and increases speed constantly by 20 m/s². Determine the relative acceleration of the second stage, A, with respect to the falling first stage, B, after the second stage has moved $x_1 = 60$ m horizontally.

We need to find the tangential and normal components of acceleration for the second stage. The tangential acceleration is equal to the increase in speed, which is constant. The normal acceleration requires us to find the speed at this location (which requires us to find the distance that rocket has traveled along the curve) and the radius of curvature at the instant desired. To find both the distance along the curve and the radius of curvature we'll need to use calculus

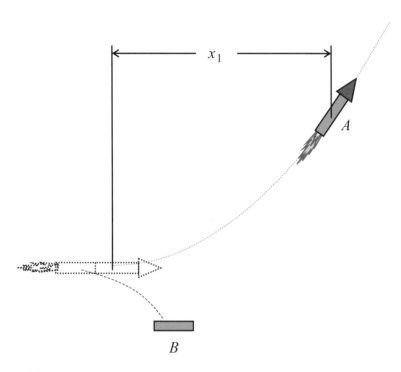

Figure 4.15: The two stage rocket of Example 4.4.

techniques from resources such as our calculus textbook or a trustworthy internet website such as Wolfram.

The length along the curve is found from https://mathworld.wolfram.com/ArcLength.html to be:

$$s = \int_{x_0}^{x_1} \sqrt{1 + (\frac{dy}{dx})^2} \, dx$$

$$\frac{dy}{dx} = 0.2x$$

$$s = \int_{0}^{60} \sqrt{1 + (0.2x)^2} \, dx.$$

From here we can solve this using integral tables, plug it into our calculator or use Wolfram Alpha. We type in "integral of sqrt(1+(0.2*x)^2)dx from x = 0 to 60" into https://www.wolframalpha.com/ which returns the answer: $s = 369.2$ m.

The speed of rocket A can be found from:

$$v_1^2 = v_0^2 + 2as$$

$$v_1 = \sqrt{(100)^2 + 2(20)(369.2)} = 157.4 \text{ m/s}.$$

The angle tangent at this location is found from the inverse tangent of the slope:

$$\frac{dy}{dx} = 0.2x = 0.2(60) = 12$$

$$\theta = \tan^{-1}(12) = 85.24°.$$

We note that Figure 4.15 is not to scale.

The radius of curvature is found from
https://mathworld.wolfram.com/RadiusofCurvature.html to be:

$$\rho = \frac{\left[1 + \left(\dfrac{dy}{dx}\right)^2\right]^{3/2}}{\left|\dfrac{d^2 y}{dx^2}\right|}.$$

Plugging in $\frac{dy}{dx} = 0.2x = 0.2(60) = 12$ and $\frac{d^2 y}{dx^2} = 0.2$:

$$\rho = \frac{\left[1 + \left(\dfrac{dy}{dx}\right)^2\right]^{3/2}}{\left|\dfrac{d^2 y}{dx^2}\right|} = \frac{\left[1 + (12)^2\right]^{3/2}}{|0.2|} = 8{,}730 \text{ m}.$$

The normal acceleration of rocket A is:

$$a_n = \frac{v^2}{\rho} = \frac{(157.4)^2}{8{,}730} = 2.838 \text{ m/s}^2.$$

The acceleration of rocket A in path coordinates is:

$$\vec{a}_A = (20)\hat{e}_t + (2.838)\hat{e}_n \text{ m/s}^2.$$

Figure 4.16 shows the overlaid unit vectors of rocket A's path and rectangular coordinates. Table 4.5 is the corresponding transformation matrix.

Applying the coordinate transformation array:

$$\vec{a}_A = a_t[\cos\theta\hat{i} + \sin\theta\hat{j}] + a_n[-\sin\theta\hat{i} + \cos\theta\hat{j}]$$

$$\vec{a}_A = (20)[\cos(85.24°)\hat{i} + \sin(85.24°)\hat{j}] + (2.838)[-\sin(85.24°\hat{i} + \cos(85.24°)\hat{j}]$$

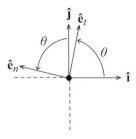

Figure 4.16: Unit vectors of Example 4.4.

Table 4.5: Coordinate transformation array of Example 4.4

	$\hat{\imath}$	$\hat{\jmath}$
$\vec{e}_t$	$\cos \theta$	$-\sin \theta$
$\vec{e}_n$	$\sin \theta$	$\cos \theta$

$$\vec{a}_A = (-1.169)\hat{\imath} + (20.17)\hat{\jmath} \text{ m/s}^2.$$

The acceleration of the first stage, assuming no aerodynamic drag, is entirely gravity:

$$\vec{a}_B = (0)\hat{\imath} + (-9.81)\hat{\jmath} \text{ m/s}^2.$$

The relative acceleration of A with respect to B is:

$$\vec{a}_{A/B} = \vec{a}_A - \vec{a}_B$$

$$\vec{a}_{A/B} = [(-1.169) - (0)]\hat{\imath} + [(20.17) - (-9.81)]\,\hat{\jmath}.$$

The answer is:

$$\boxed{\vec{a}_{A/B} = (-1.17))\hat{\imath} + (30.0)\hat{\jmath} \text{ m/s}^2}.$$

Book 1 - Class 5

https://www.youtube.com/watch?v=k6FBk-Fw_0k

Non-Rectangular Coordinate Systems: Polar Coordinates

B.L.U.F. (Bottom Line Up Front)

- Polar Coordinates (a.k.a. "Radial and Transverse"): rotating about an origin to follow a particle, based on distance from the origin and angle from horizontal.

- Velocity is expressed in Polar Coordinates as: $\vec{\mathbf{v}} = (\dot{r})\hat{\mathbf{e}}_r + (r\dot{\theta})\hat{\mathbf{e}}_\theta$.

- Acceleration is expressed in Polar Coordinates as: $\vec{\mathbf{a}} = (\ddot{r} - r\dot{\theta}^2)\hat{\mathbf{e}}_r + (r\ddot{\theta} + 2\dot{r}\dot{\theta})\hat{\mathbf{e}}_\theta$.

5.1 POLAR COORDINATES (A.K.A. "RADIAL AND TRANSVERSE")

There are situations where a fixed point is a convenient reference from which to track a particle's movement. One such instance is the use of radar (which is an acronym for "radio detection and ranging") that can find the direction and distance of an object from a reference location.

The cartoon in Figure 5.1 is intended to help visualize polar coordinates. Newtdog is sitting on a spinning stool and holding Wormy on the end of a fishing rod line that can spool outward. The velocity and acceleration of Wormy can be described using both the fishing line and the spinning stool. The length of fishing line can increase or decrease, so the radial distance from the center of rotation has a rate of change we can call $\dot{r}$ and $\ddot{r}$. The angular position of the stool changes with time and that rate can can be described as $\dot{\theta}$ and $\ddot{\theta}$. These parameters are used together to fully describe Wormy's motion.

Note that the same motion can also be described in Rectangular or Path coordinate systems. It's just that this scenario is well suited for polar coordinates since the arrangement is most easily described using distance and angle.

As with path coordinates, we introduce the equations and conventions and then describe how they came about. Figure 5.2 shows a particle following a path and the position vector. The radial direction unit vector $\hat{\mathbf{e}}_r$ is aligned with the position vector, starting from reference point O and going to point P. The transverse direction unit vector $\hat{\mathbf{e}}_\theta$ is perpendicular to $\hat{\mathbf{e}}_r$ and by

Figure 5.1: Newtdog demonstrating polar coordinates with Wormy on a fishing line (repeat of Figure 4.2) (©E. Diehl).

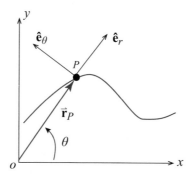

Figure 5.2: Polar coordinates unit vectors.

the right-hand rule convention goes to the left, or counter-clockwise, from the radial direction. This is because counter-clockwise rotation is treated as positive and clockwise as negative. It's important to note that $\hat{\mathbf{e}}_r$ and $\hat{\mathbf{e}}_\theta$ are *not* tangent or perpendicular to the path (unless the path is a perfect circle), as shown in Figure 5.3.

The position vector(s) *could* be written as $\vec{\mathbf{r}}_P = r\hat{\mathbf{e}}_r$ and $\vec{\theta} = \theta\hat{\mathbf{e}}_\theta$ but the latter isn't needed as the angle for the position is established by the radial direction position vector.

The velocity vector of a particle in polar coordinates is

$$\boxed{\vec{\mathbf{v}} = \left(\dot{r}\right)\hat{\mathbf{e}}_r + \left(r\dot{\theta}\right)\hat{\mathbf{e}}_\theta} \quad \text{or} \quad \boxed{\vec{\mathbf{v}} = v_r\hat{\mathbf{e}}_r + v_\theta\hat{\mathbf{e}}_\theta}.$$

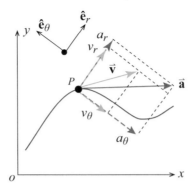

Figure 5.3: Velocity and acceleration components in path coordinates.

The acceleration vector of a particle in polar coordinates is

$$\vec{a} = \left(\ddot{r} - r\dot{\theta}^2\right)\hat{e}_r + \left(r\ddot{\theta} + 2\dot{r}\dot{\theta}\right)\hat{e}_\theta \quad \text{or} \quad \vec{a} = a_r\hat{e}_r + a_\theta\hat{e}_\theta \,,$$

where the *variables of interest* (V.o.I.) are:

r = radius of position vector from origin to the particle

$\dot{r}$ = rate of change of the radius

$\ddot{r}$ = rate of change of the rate of change of the radius

θ = angular position of the position vector with respect to horizontal counter-clockwise

$\dot{\theta}$ = rate of change of the angular position (units of rad/s, but also rpm)

$\ddot{\theta}$ = rate of change of the rate of change of the angular position (units of rad/s²)

In many problems using polar coordinates, a good place to start is to identify the known and unknown V.o.I. so you can gather them to find the components of polar coordinates which are:

$$v_r = \dot{r} \quad \text{and} \quad v_\theta = r\dot{\theta}$$
$$a_r = \ddot{r} - r\dot{\theta}^2 \quad \text{and} \quad a_\theta = r\ddot{\theta} + 2\dot{r}\dot{\theta}.$$

A few things to emphasize with polar coordinates include: (a) the acceleration components are NOT the derivatives of the velocity components (i.e., $a_r \neq \frac{d(v_r)}{dt}$ and $a_\theta \neq \frac{d(v_\theta)}{dt}$); (b) $\hat{e}_\theta$ always goes to the left (counter-clockwise) of $\hat{e}_r$ which is not true of the normal and tangential

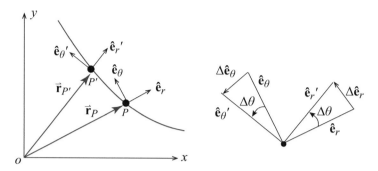

Figure 5.4: Consequence of moving unit vectors in polar coordinates.

unit vectors since the normal direction depends on the path curvature; and (c) the magnitude of velocity (speed) is found from $|v| = \sqrt{v_r{}^2 + v_\theta{}^2}$ and the magnitude of acceleration from $|a| = \sqrt{a_r{}^2 + a_\theta{}^2}$.

With introduction of the equations out of the way, we can mathematically investigate where the velocity and acceleration components come from. We begin this time with the position vector: $\vec{\mathbf{r}} = r\hat{\mathbf{e}}_r$. The velocity is the time rate change of this position vector so using the product rule again we get:

$$\vec{\mathbf{v}} = \frac{d\,(r\hat{\mathbf{e}}_r)}{dt} = \frac{dr}{dt}\hat{\mathbf{e}}_r + r\frac{d\,(\hat{\mathbf{e}}_r)}{dt}.$$

Similar to what was done in path coordinates, we need to find $\frac{d(\hat{\mathbf{e}}_r)}{dt}$. We again pick two points on a path and examine the changes in the unit vectors. This time we'll pick a path and two points where the rotation is counter-clockwise so it will be positive.

Figure 5.4 tracks the motion from point P to point P' their position vectors and associated unit vectors. While the unit vectors are shown attached, they can be moved on top of each other as in the right portion of the figure. When unit vector triangles are drawn, we observe that $\hat{\mathbf{e}}_r' = \hat{\mathbf{e}}_r + \Delta\hat{\mathbf{e}}_r$ and $\hat{\mathbf{e}}_\theta' = \hat{\mathbf{e}}_\theta + \Delta\hat{\mathbf{e}}_\theta$. As points P get closer P', the Δ's get smaller. We note the directions become:

$$\lim_{\Delta\to 0}\frac{\Delta\hat{\mathbf{e}}_r}{\Delta\theta} = \frac{d\,(\hat{\mathbf{e}}_r)}{d\theta} = \hat{\mathbf{e}}_\theta \quad \text{and} \quad \lim_{\Delta\to 0}\frac{\Delta\hat{\mathbf{e}}_\theta}{\Delta\theta} = \frac{d\,(\hat{\mathbf{e}}_\theta)}{d\theta} = -\hat{\mathbf{e}}_r.$$

This is an interesting result in that the changing radial unit vector aligns with the transverse unit vector, BUT the changing transverse unit vector aligns with the radial unit vector in the NEGATIVE direction.

This result is for unit vectors changing with an angular change, but we want to know how they change with time, so we again use the chain rule, and note that $\frac{d\theta}{dt} = \dot{\theta}$:

$$\frac{d\,(\hat{\mathbf{e}}_r)}{dt} = \frac{d\,(\hat{\mathbf{e}}_r)}{d\theta}\frac{d\theta}{dt} = \dot{\theta}\hat{\mathbf{e}}_\theta \quad \text{and} \quad \frac{d\,(\hat{\mathbf{e}}_\theta)}{dt} = \frac{d\,(\hat{\mathbf{e}}_\theta)}{d\theta}\frac{d\theta}{dt} = -\dot{\theta}\hat{\mathbf{e}}_r.$$

Going back to the velocity calculation we can use the first result to find $\vec{v} = \frac{dr}{dt}\hat{e}_r + r\frac{d(\hat{e}_r)}{dt} = \dot{r}\hat{e}_r + r\dot{\theta}\hat{e}_\theta$, giving us the velocity equation in polar coordinates:

$$\vec{v} = (\dot{r})\,\hat{e}_r + \left(r\dot{\theta}\right)\hat{e}_\theta\,.$$

To find the acceleration we differentiate velocity with respect to time, switch to dot notation, and get:

$$\vec{a} = \frac{d\vec{v}}{dt} = \ddot{r}\hat{e}_r + \dot{r}\underbrace{\dot{\hat{e}}_r}_{\dot{\theta}\hat{e}_\theta} + \dot{r}\dot{\theta}\hat{e}_\theta + r\ddot{\theta}\hat{e}_\theta + r\dot{\theta}\underbrace{\dot{\hat{e}}_\theta}_{-\dot{\theta}\hat{e}_r}\,.$$

Note that because there are three terms in the $\hat{e}_\theta$ part of the velocity we must use the product rule three times. Also note that $\dot{r}\dot{\theta}\hat{e}_\theta$ appears twice. Gathering the terms we find the acceleration equation in polar coordinates:

$$\vec{a} = \left(\ddot{r} - r\dot{\theta}^2\right)\hat{e}_r + \left(r\ddot{\theta} + 2\dot{r}\dot{\theta}\right)\hat{e}_\theta\,.$$

We should keep in mind why there are multiple parts to this equation: the unit vectors change position with time.

Within this equation are noteworthy parts. First, note that if the radius were constant (which is circular motion where: $\dot{r} = 0$ and $\ddot{r} = 0$), then the radial component of acceleration is $\left(-r\dot{\theta}^2\right)\hat{e}_r$. In this circular motion the velocity is only in the transverse direction and the speed is $v = r\dot{\theta}$. Next, we note that by rearranging $\dot{\theta} = \frac{v}{r}$ and substituting into the radial component of acceleration we get $-r\dot{\theta}^2\hat{e}_r = -\frac{v^2}{r}\hat{e}_r$. We should recognize this as the normal component of path coordinates, $\left(\frac{v^2}{\rho}\right)\hat{e}_n$. We also note that in the case of circular motion $\hat{e}_n = -\hat{e}_r$. It's useful to identify this term as "centrifugal."

Another term of interest is $2\dot{r}\dot{\theta}$ which is associated with "Coriolis Acceleration." This is a particular kind of acceleration that only exists when there is radial velocity and angular velocity. It will be discussed further in rigid body kinematics in Class 18 (vol. 2), but it's worth noting its significance here. The Coriolis Acceleration phenomenon can be felt if one walks outward on a rotating platform such as a playground merry-go-round. You can even feel the mysterious force that pushes you in an odd direction.

The Coriolis Acceleration phenomenon even controls some weather patterns due to the Earth's rotational speed, notably hurricanes which rotate counter-clockwise in the Northern Hemisphere and clockwise in the Southern. A popular myth says Coriolis also controls the direction water drains in sinks and toilet bowls, but the Earth's relatively slow rotational speed doesn't create enough of this acceleration to have this effect. Also, eggs don't balance any better during the vernal equinox than any other day, and glass is not a very viscous liquid that continues to flow.

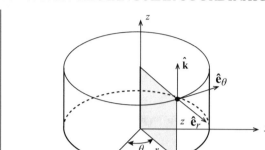
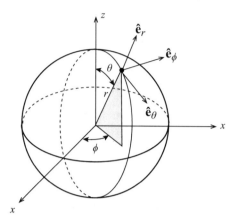

Figure 5.5: Cylindrical and spherical coordinates.

5.2   POLAR COORDINATES EXTENDED TO THREE DIMENSIONS

Polar coordinates are the two-dimensional version of "cylindrical coordinates" which have a third unit vector out of the page, obeying the right-hand rule as shown in the left image in Figure 5.5, where the thumb is $\hat{\mathbf{e}}_r$, index finger is $\hat{\mathbf{e}}_\theta$, and middle finger is $\hat{\mathbf{k}}$. Describing helical motion is an example where cylindrical coordinates work well. We will need a third coordinate to describe rotation vectors when dealing with rigid body motion.

Another related coordinate system is spherical coordinates as shown in the image to the right of Figure 5.5. The third coordinate in spherical coordinates is an additional angle represented by the unit vector $\hat{\mathbf{e}}_\phi$. Spherical coordinates are useful for describing orbital motion or for radar tracking that isn't limited to one plane.

The important take-away in this class topic is: any coordinate system can be used, but some work better with particular situations.

Example 5.1

A new design for an extendable crane (with the operator positioned at the end as shown in Figure 5.6) rotates so $\theta = (2/\pi)\sin(\pi t)$ rad and extends with $r = 3\left(2 + 2e^{-1.5t}\right)$ m. When $t = 2.25$ s, determine: (a) the velocity of point A in polar coordinates ($\hat{\mathbf{e}}_r$ and $\hat{\mathbf{e}}_\theta$); (b) the velocity of point A in rectangular coordinates ($\hat{\imath}$ and $\hat{\jmath}$); (c) the acceleration of point A in polar coordinates ($\hat{\mathbf{e}}_r$ and $\hat{\mathbf{e}}_\theta$); and (d) the acceleration of point A rectangular coordinates ($\hat{\imath}$ and $\hat{\jmath}$). Plot the path of this motion.

When we recognize a problem should use polar coordinates, looking for the V.o.I. is a good first step.

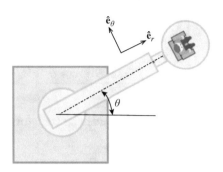

Figure 5.6: Example 5.1 extendable crane.

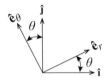

Figure 5.7: Unit vectors of Example 5.1.

V.o.I.:

$$r = 3\left(2 + 2e^{-1.5t}\right) \qquad \theta = (2/\pi)\sin(\pi t)$$
$$\dot{r} = -9e^{-1.5t} \qquad \dot{\theta} = (2)\cos(\pi t)$$
$$\ddot{r} = 13.5e^{-1.5t} \qquad \ddot{\theta} = -(2\pi)\sin(\pi t)$$

@ $t = 2.25$ s:

$$r = 6.205 \text{ m} \qquad \theta = 0.4502 \text{ rad } (25.79°)$$
$$\dot{r} = -0.3080 \text{ m/s} \qquad \dot{\theta} = 1.414 \text{ rad/s}$$
$$\ddot{r} = 0.4619 \text{ m/s}^2 \qquad \ddot{\theta} = 4.443 \text{ rad/s}^2.$$

Find the velocity

$$\vec{\mathbf{v}}_A = (\dot{r})\,\hat{\mathbf{e}}_r + \left(r\dot{\theta}\right)\hat{\mathbf{e}}_\theta = (-0.3080)\,\hat{\mathbf{e}}_r + ((6.205)(1.414)\,)\,\hat{\mathbf{e}}_\theta$$
$$= (-0.3080)\,\hat{\mathbf{e}}_r + (8.776)\,\hat{\mathbf{e}}_\theta \text{ m/s}.$$

(a) $\boxed{\vec{\mathbf{v}}_A = (-0.308)\,\hat{\mathbf{e}}_r + (8.78)\,\hat{\mathbf{e}}_\theta \text{ m/s}}$

Table 5.1: Coordinate transformation matrix for Example 5.1

	$\hat{\imath}$	$\hat{\jmath}$
$\hat{e}_r$	$\cos\theta$	$\sin\theta$
$\hat{e}_\theta$	$-\sin\theta$	$\cos\theta$

Using the unit vectors of Figure 5.7 and coordinate transformation in Table 5.1 we change from polar to rectangular coordinates.

$$\vec{v}_A \left[(-0.3080)\cos(25.79°) - (8.776)\sin(25.79°)\right]\hat{\imath}$$
$$+ \left[(-0.3080)\sin(25.79°) + (8.776)\cos(25.79°)\right]\hat{\jmath}$$

$$\vec{v}_A = (-4.096)\hat{\imath} + (7.767)\hat{\jmath} \text{ m/s}$$

(b) $\boxed{\vec{v}_A = (-4.10)\hat{\imath} + (7.77)\hat{\jmath} \text{ m/s}}$.

Find the acceleration

$$\vec{a}_A = \left(\ddot{r} - r\dot{\theta}^2\right)\hat{e}_r + \left(r\ddot{\theta} + 2\dot{r}\dot{\theta}\right)\hat{e}_\theta$$
$$= \left((0.4619) - (6.205)(1.414)^2\right)\hat{e}_r + ((6.205)(4.443)$$
$$+ 2(-0.3080)(1.414))\hat{e}_\theta \text{ m/s}^2$$

$$\vec{a}_A = (-11.95)\hat{e}_r + (26.70)\hat{e}_\theta \text{ m/s}^2$$

(c) $\boxed{\vec{a}_A = (-12.0)\hat{e}_+ (26.7)\hat{e}_\theta \text{ m/s}^2}$

$$\vec{a}_A = \left[(-11.95)\cos(25.79°) + (26.70)\sin(25.79°)\right]\hat{\imath}$$
$$+ \left[(-11.95)\sin(25.79°) + (16.70)\cos(25.79°)\right]\hat{\jmath} \text{ m/s}^2$$

$$\vec{a}_A = (-22.38)\hat{\imath} + (18.84)\hat{\jmath} \text{ m/s}^2$$

(d) $\boxed{\vec{a}_A = (-22.4)\hat{\imath} + (18.8)\hat{\jmath} \text{ m/s}^2}$.

What does this motion look like? Figure 5.8 shows a graph of the motion described by the above. We note that the arm starts on the horizontal axis at 12 m and moves to the orange square after 2.25 s along the path shown. The graph helps confirm the velocity as it is tangent, that is upward and to the left based on the results in part (b).

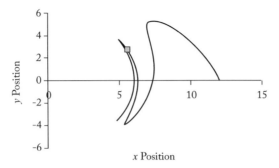

Figure 5.8: Graphs of motion in Example 5.1.

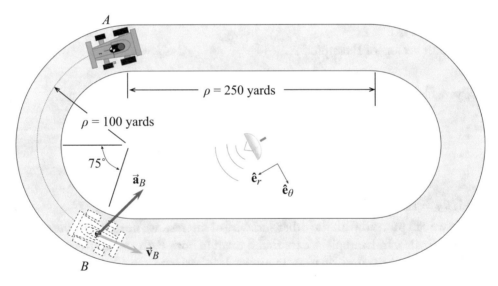

Figure 5.9: Example 5.4 race track.

Example 5.2

The race car from Example 4.2 is being tracked by a radar unit positioned in the center of the track which uses polar coordinates. Given the location of the radar as shown in Figure 5.9, determine the radial and angular (transverse) parameters (variables of interest: $r, \dot{r}, \ddot{r}, \theta, \dot{\theta}, \ddot{\theta}$) it will report when the car is at point B.

Table 5.2: Coordinate transformation of Example 5.2

	$\hat{\mathbf{i}}$	$\hat{\mathbf{j}}$
$\hat{\mathbf{e}}_r$	$-\cos\theta$	$-\sin\theta$
$\hat{\mathbf{e}}_\theta$	$\sin\theta$	$-\cos\theta$

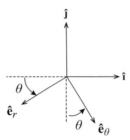

Figure 5.10: Unit vectors of Example 5.2.

The variables of interest begin with some geometry:

$$r = \sqrt{\left(\underbrace{300\sin 75°}_{289.8}\right)^2 + \left(\underbrace{375 + 300\cos 75°}_{452.7}\right)^2} = 537.5 \text{ ft.}$$

The angle is $\theta_B = \tan^{-1}\left(\frac{(289.8)}{(452.7)}\right) = 32.63°$ clockwise from the horizontal.

Before we can proceed with the other variables of interest, we need to put the acceleration into polar coordinates. In Example 4.2 we found them in both Path and Cartesian coordinates, so we could perform the transformation in either. It is easier to use Cartesian. The results of Example 4.2 were:

$$\vec{v} = (354.2)\,\hat{\imath} - (94.91)\hat{\jmath} \text{ ft/s}^2$$

$$\vec{a} = (185.4)\,\hat{\imath} + (414.4)\hat{\jmath} \text{ ft/s}^2.$$

With the unit vectors shown in Figure 5.10, we generate the transformation array in Table 5.2. We use the column direction to get the vectors into Polar:

$$\vec{v} = (354.3)\left[-\cos\theta\hat{\mathbf{e}}_r + \sin\theta\hat{\mathbf{e}}_\theta\right] + (-94.91)\left[-\sin\theta\hat{\mathbf{e}}_r - \cos\theta\hat{\mathbf{e}}_\theta\right]$$

$$\vec{v} = \left[-(354.2)\cos(32.63°) - (-94.91)\sin(32.63°)\right]\hat{\mathbf{e}}_r$$
$$+ \left[(354.2)\sin(32.63°) - (-94.91)\cos(32.63°)\right]\hat{\mathbf{e}}_\theta$$

$$\vec{v} = \underbrace{(-247.1)}_{v_r} \hat{e}_r + \underbrace{(270.9)}_{v_\theta} \hat{e}_\theta \text{ ft/s}$$

$$\vec{a} = (185.4)\left[-\cos\theta\hat{e}_r + \sin\theta\hat{e}_\theta\right] + (414.4)\left[-\sin\theta\hat{e}_r - \cos\theta\hat{e}_\theta\right]$$

$$\vec{a} = \left[-(185.4)\cos\left(32.63°\right) - (414.4)\sin\left(32.63°\right)\right]\hat{e}_r$$
$$+ \left[(185.4)\sin\left(32.63°\right) - (414.4)\cos\left(32.63°\right)\right]\hat{e}_\theta$$

$$\vec{a} = \underbrace{(-378.6)}_{a_r} \hat{e}_r + \underbrace{(-249.0)}_{a_\theta} \hat{e}_\theta \text{ ft/s}^2.$$

We can now find the variables of interest.

We begin with $\dot{r}$ which is the speed the car is moving toward the radar at this instant and equal to the radial component of velocity:

$$\dot{r} = v_r = -274.1 \text{ ft/s}.$$

Next, we use v_θ to find $\dot{\theta}$, which is the rotational speed of the radar at this instant if it were a rotating dish following the car, which isn't really how radar tracks things, but it's a good visual:

$$v_\theta = r\dot{\theta} = 270.9 \text{ ft/s}$$

$$\dot{\theta} = \frac{v_\theta}{r} = \frac{(270.9)}{(537.5)} = 0.5040 \text{ rad/s}.$$

The acceleration terms require a bit more effort. For the radial acceleration, we use the angular speed we just found to isolate it from the radial component:

$$a_r = \ddot{r} - r\dot{\theta}^2 = -379.6 \text{ ft/s}^2$$

$$\ddot{r} = a_r + r\dot{\theta}^2 = (-379.6) + (537.5)(0.5040)^2 = -243.1 \text{ ft/s}^2$$

$$a_\theta = r\ddot{\theta} + 2\dot{r}\dot{\theta} = -249.0 \text{ ft/s}^2$$

$$\ddot{\theta} = \frac{a_\theta - 2\dot{r}\dot{\theta}}{r} = \frac{(-249.0) - 2(-274.1)(0.5040)}{(537.5)} = 3.289E - 4 \text{ ft/s}^2.$$

So the V.o.I. are:

$$\boxed{r = 538 \text{ ft}} \qquad \boxed{\dot{r} = -274 \text{ ft/s}} \qquad \boxed{\ddot{r} = -243 \text{ ft/s}^2}$$

$$\boxed{\theta = 32.6° \nearrow} \qquad \boxed{\dot{\theta} = 0.504 \text{ rad/s}} \qquad \boxed{\ddot{\theta} = 3.29E - 4 \text{ rad/s}^2}.$$

Example 5.3   A cam is a machined piece that controls a desired motion, usually having another piece in contact with and following along a curved edge. In Figure 5.11 a small ball (treated as

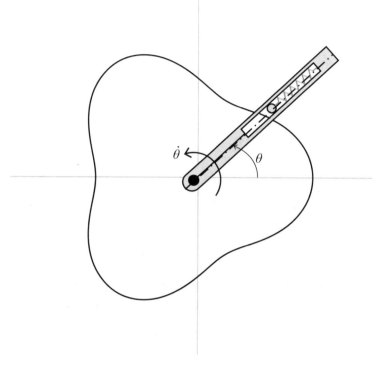

Figure 5.11: Cam mechanism of Example 5.3.

a particle) is constrained within a bar that rotates while the three lobed cam is stationary. A spring within the follower holds it against the cam which has a shaped described by $r(\theta) = 2.5 + 0.5 \cos 3\theta$ in. The follower bar rotates at a constant 150 rpm. Determine the acceleration in polar, path and Cartesian coordinates when $\theta = 105°$.

We begin with the V.o.I. We already know the angles:

$$\theta = 105°$$

$$\dot{\theta} = (100 \text{ rpm})\frac{(2\pi \text{ rad/rev})}{(60 \text{ s/min})} = 15.71 \text{ rad/s}.$$

$$\ddot{\theta} = 0.$$

The radial V.o.I. require that we take the time derivative of the given shape function, but we note it is a function of angular position rather than time. We need to use the chain rule to account for this $\frac{dr(\theta)}{dt} = \frac{dr(\theta)}{d\theta}\frac{d\theta}{dt} = \dot{\theta}\frac{dr(\theta)}{d\theta}$:

$$r = 2.5 + 0.5 \cos 3\theta = 2.5 + 0.5 \cos(3(105°)) = 2.854 \text{ in.}$$

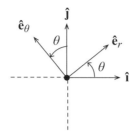

Figure 5.12: Unit vectors for Polar coordinate transformation of Example 5.3.

Table 5.3: Polar coordinate transformation matrix of Example 5.3

	$\hat{\imath}$	$\hat{\jmath}$
$\hat{\mathbf{e}}_r$	$\cos\theta$	$\sin\theta$
$\hat{\mathbf{e}}_\theta$	$-\sin\theta$	$\cos\theta$

$$\dot{r} = -1.5\dot{\theta}\sin 3\theta = -1.5(15.71)\sin(3(105°)) = 16.66 \text{ in/s.}$$

$$\ddot{r} = -1.5\ddot{\theta}\sin 3\theta - 4.5\dot{\theta}^2\cos 3\theta$$
$$= -1.5(0)\sin(3(105°)) - 4.5(15.71)^2\cos(3(105°)) = -785.3 \text{ in/s}^2.$$

Note for the second time derivative above it is necessary to use the product rule.

The acceleration in polar coordinates is:

$$\vec{a} = \left(\ddot{r} - r\dot{\theta}^2\right)\hat{\mathbf{e}}_r + \left(r\ddot{\theta} + 2\dot{r}\dot{\theta}\right)\hat{\mathbf{e}}_\theta$$

$$\vec{a} = ((-785.3) - (2.854)(15.71)^2)\hat{\mathbf{e}}_r + ((2.854)(0) + 2(16.66)(15.71))\hat{\mathbf{e}}_\theta \text{ in/s}^2$$

$$\vec{a} = (-1{,}490)\hat{\mathbf{e}}_r + (523.5)\hat{\mathbf{e}}_\theta \text{ in/s}^2.$$

The coordinate transformation into Cartesian coordinates is done using the unit vectors shown in Figure 5.12 and the resulting transformation matrix in Table 5.3. To avoid some confusion, the unit vectors overlaid are drawn with an acute angle even though the actual angle is obtuse:

$$\vec{a} = [(-1{,}490)\cos(105°) - (523.5)\sin(105°)]\hat{\imath}$$
$$+ [(-1{,}490)\sin(105°) + (523.5)\cos(105°)]\hat{\jmath} \text{ in/s}^2$$

$$\vec{a} = (-120.0)\hat{\imath} + (-1{,}575)\hat{\jmath} \text{ in/s}^2.$$

In order to transform the acceleration into path coordinates, we need to know the direction of the tangential component (and from that we can find the direction of the normal component).

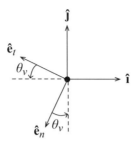

Figure 5.13: Unit vectors for path coordinate transformation of Example 5.3.

Table 5.4: Path coordinate transformation matrix of Example 5.3

	$\hat{\imath}$	$\hat{\jmath}$
$\hat{\mathbf{e}}_t$	$-\cos\theta_v$	$\sin\theta_v$
$\hat{\mathbf{e}}_n$	$-\sin\theta_v$	$-\cos\theta_v$

We know that velocity is tangent to the path, so we use that knowledge, transform the velocity in polar coordinates into rectangular coordinates and from that find the angle of the velocity vector:

$$\vec{\mathbf{v}} = \dot{r}\hat{\mathbf{e}}_r + r\dot{\theta}\hat{\mathbf{e}}_\theta = (16.66)\hat{\mathbf{e}}_r + ((2.854)(15.71))\hat{\mathbf{e}}_\theta$$

$$\vec{\mathbf{v}} = (16.66)\hat{\mathbf{e}}_r + (44.84)\hat{\mathbf{e}}_\theta \text{ in/s}$$

$$\vec{\mathbf{v}} = [(16.66)\cos(105°) - (44.84)\sin(105°)]\hat{\imath}$$
$$+ [(16.66)\sin 105°) + (44.84)\cos(105°)]\hat{\jmath} \text{ in/s}$$

$$\vec{\mathbf{v}} = (-47.62)\hat{\imath} + (4.487)\hat{\jmath} \text{ in/s}.$$

The velocity magnitude and direction are:

$$v = \sqrt{(-47.62)^2 + (4.487)^2} = 47.83 \text{ in/s}$$

$$\theta_v = \tan\frac{(4.487)}{(47.62)} = 5.383° \searrow.$$

The coordinate transformation into path coordinates is done using the unit vectors shown in Figure 5.13 and the resulting transformation matrix in Table 5.4:

$$\vec{\mathbf{a}} = [-(-120.0)\cos(5.383°) + (-1{,}575)\sin(5.383°)]\hat{\mathbf{e}}_t$$
$$+ [-(-120.0)\sin(5.383°) - (-1{,}575)\cos(5.383°)]\hat{\mathbf{e}}_n \text{ in/s}^2$$

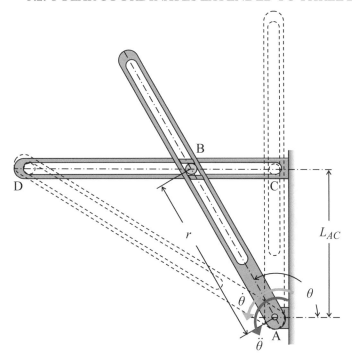

Figure 5.14: Mechanism of Example 5.4.

$$\vec{a} = (-1{,}552)\hat{e}_t + (-291.7)\hat{e}_n \text{ in/s}^2.$$

Answers:

Polar coordinates: $\boxed{\vec{a} = (-1{,}490)\hat{e}_r + (523.5)\hat{e}_\theta \text{ in/s}^2}$

Path coordinates: $\boxed{\vec{a} = (-1{,}552)\hat{e}_t + (-291.7)\hat{e}_n \text{ in/s}^2}$

Rectangular coordinates: $\boxed{\vec{a} = (-120.0)\hat{i} + (-1{,}575)\hat{j} \text{ in/s}^2}$

Example 5.4   The mechanism in Figure 5.14 has a small ball (treated as particle B) constrained between two slotted bars, one stationary and the other rotating at 3 rad/s and 1 rad/s^2 (both counter-clockwise) at the instant shown. The bases of the bars are $L_{AC} = 0.5$ m apart. Determine the acceleration of the particle at the instant shown ($\theta = 120°$).

This is a type of dependent motion problem somewhat different than what was introduced in Class 2. In this case we recognize that the radius, r, is dependent on the angle, θ. We can find this relationship from simple trigonometry:

$$L_{AC} = r \sin \theta$$

$$r(\theta) = \frac{L_{AC}}{\sin\theta} = L_{AC}\csc\theta$$

$$r(\theta) = (0.5)\csc(120°) = 0.5774 \text{ m}.$$

To find the radial variables of interest we take the time derivative of this relationship:

$$\dot{r} = -L_{AC}\dot{\theta}\csc\theta\cot\theta = -(0.5)(3)\csc(120°)\cot(120°) = -1.732 \text{ m/s}.$$

To find the radial acceleration, we need to remember to take the time derivative of each part based on the product rule:

$$\ddot{r} = -L_{AC}\left[\ddot{\theta}\csc\theta\cot\theta - \dot{\theta}^2\csc\theta\cot\theta\cot\theta - \dot{\theta}^2\csc\theta\csc^2\theta\right]$$

$$\ddot{r} = -L_{AC}\left[\ddot{\theta}\csc\theta\cot\theta - \dot{\theta}^2\csc\theta\cot^2\theta - \dot{\theta}^2\csc^3\theta\right]$$

$$\ddot{r} = -(0.5)\left[(1)\csc(120°)\cot(120°) - (3)^2\csc(120°)\cot^2(120°)\right.$$
$$\left. -(3)^2\csc^3(120°)\right] = 38.42 \text{ m/s}^2.$$

The acceleration in polar coordinates is:

$$\vec{a} = \left(\ddot{r} - r\dot{\theta}^2\right)\hat{e}_r + \left(r\ddot{\theta} + 2\dot{r}\dot{\theta}\right)\hat{e}_\theta$$

$$\vec{a} = ((38.42) - (0.5)(3)^2)\hat{e}_r + ((0.5)(1) + 2(-1.732)(3))\hat{e}_\theta \text{ m/s}^2$$

$$\vec{a} = ((33.92)\hat{e}_r + (-9.892)\hat{e}_\theta \text{ m/s}^2.$$

Answer:

$$\boxed{\vec{a} = (33.9)\hat{e}_r + (-9.89)\hat{e}_\theta \text{ m/s}^2}.$$

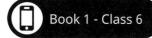

https://www.youtube.com/watch?v=XZ9wi8jwe5c

C L A S S   6

Newton's Second Law (N2L) in Rectangular Coordinates

B.L.U.F. (Bottom Line Up Front)

- Kinetics: Newton's Second Law: $\sum \vec{\mathbf{F}} = m\,\vec{\mathbf{a}}$, net force is proportional to acceleration.

- Rectangular coordinates: $\rightarrow \sum F_x = ma_x$ and $\uparrow \sum F_y = ma_y$.

6.1 KINETICS VS. KINEMATICS

It's worth repeating the definitions of the two basic Dynamics topics: kinematics is the description of motion; and kinetics is the reason the motion is happening. In kinetics we introduce cause and effect. We will use three separate kinetics analysis techniques: Newton's Second Law (N2L), Work-Energy, and Impulse-Momentum. We'll see that these three are closely related and often all three can be used along with kinematics to get the same results.

6.2 NEWTON'S SECOND LAW

In 1687, Sir Isaac Newton published *Philosophiæ Naturalis Principia Mathematica* (Mathematical Principles of Natural Philosophy) with his three laws of motion. You've covered the first and third laws in Statics class, but in Dynamics we also use Newton's second law, using the shorthand "N2L" because it sounds cool. Newton wrote his famous paper in Latin, so translations into English can vary, plus there are several physical implications, so it isn't surprising that N2L can be written several ways. Below are two useful ones along with their associated equations.

- The most famous: $\boxed{\sum \vec{\mathbf{F}} = m\,\vec{\mathbf{a}}}$ which means "If the resultant force acting on a particle is not zero, the particle will have an acceleration proportional to the magnitude of the resultant and in the direction of the resultant force."

- Another way: $\sum \vec{\mathbf{F}} = \frac{d}{dt}\left(\vec{\mathbf{L}}\right) = \frac{d}{dt}\left(m\vec{\mathbf{v}}\right)$ which means "The second law states that the net force on a particle is equal to the time rate of change of its linear momentum $\left(\vec{\mathbf{L}} = m\vec{\mathbf{v}}\right)$ in an inertial reference frame."

Figure 6.1: Newtdog solves his homework (© E. Diehl).

In the latter case, we note that if mass is constant (which is typically the case in Dynamics, although not always in Fluid Mechanics or Thermodynamics), we arrive at the same place as the first version:

$$\sum \vec{\mathbf{F}} = \frac{d}{dt}(m\vec{\mathbf{v}}) = m\frac{d\vec{\mathbf{v}}}{dt} = m\vec{\mathbf{a}}.$$

So why bother mentioning it? As we'll see, the other two kinetics methods, Work-Energy and Impulse-Momentum, are closely related to N2L because of the "time rate change of momentum" concept.

The cartoon shown in Figure 6.1 is a humorous portrayal of Newtdog doing his homework. He's stuck while solving his most famous problem until his own "ah-ha" moment strikes him (lightbulbs aren't invented yet, so he has a "candle light up," that's why it's funny).

This is more than just a sad attempt at humor. Developing your "ah-ha'" moment skills is really important to becoming an engineer. Solving Dynamics problems is difficult and requires effort and determination. Bear in mind that Newton, while incredibly groundbreaking, was a real person like you and me, who struggled but persevered. He's quoted as saying "If I have ever made any valuable discoveries, it has been owing more to patient attention, than to any other talent."

6.3 FREE BODY DIAGRAM (FBD) AND INERTIA BODY DIAGRAM (IBD)

We'll approach kinetics in Dynamics as an extension of Statics by using its most powerful tool: the Free Body Diagram (FBD)! To do this we combine the FBD with a new diagram that represents the right side of the N2L equation, the "Inertia Body Diagram" (IBD) (Figure 6.2).

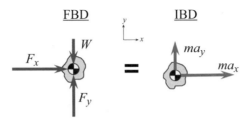

Figure 6.2: The Freed Body Diagram (FBD) and Inertial Body Diagram (IBD) pair $\sum \vec{F} = m\vec{a}$.

The rules for FBDs still apply: isolate the object (we're using particles right now, so their shape doesn't matter yet) and draw all of the EXTERNAL forces acting ON the object. If these forces balance, there is no acceleration, but if the forces don't balance, we have acceleration. To address this we include the new IBD. Some textbooks call this the "inertial response diagram" or the "kinetic diagram." Regardless of what you call it, this is a convenient way to represent N2L because it is a graphical representation of the equation. Unfortunately, there exist other conventions in various textbooks that can add some confusion to the struggling student who runs across them. You are of course encouraged to look for additional resources, but keep in mind you may run into other ways to approach N2L problems.

Take special note of some of the features of the FBD/IBD diagram pair. First, note that it matches the N2L equation below it. You should draw the FBD/IBD first thing, every time, and then write the N2L equations below it. Note that the weight (which equals mass times the acceleration of gravity) belongs on the FBD side, since gravitational force is an external force. Even though the acceleration of gravity has the word "acceleration" in it, we do not include that acceleration on the IBD. Other things to note include the coordinate axis label, the forces drawn acting on the center of mass of the particle (whose size and shape don't matter, yet), the inertial motion drawn as vectors acting from the center of mass of the particle, and especially how the forces and motion vector are proportional. This last part isn't 100% necessary in your diagrams but is done here to emphasize that force and acceleration are proportional (by mass) and the coordinate directions are independent of one another (for now). This means we can write individual equations for each direction (be sure to declare the positive direction):

$$\rightarrow \sum F_x = ma_x \quad \text{and} \quad \uparrow \sum F_y = ma_y.$$

Separating the directions like this is useful when applying N2L to other coordinate systems as we'll do in the next classes. We'll also see that kinetics can be tied to kinematics on the right hand side of the equation.

When the equations are written like this they are often referred to as the "equations of motion" which are commonly used in other courses such as vibrations and control systems.

Figure 6.3: Newtdog sets up an FBD/IBD pair (© E. Diehl).

It's best to avoid memorizing step-by-step solutions in Dynamics because there are just too many variations. But there are procedures you should follow to keep organized and avoid common mistakes. The FBD/IBD convention (Figure 6.3) is an important one to follow:

- Sketch the isolated object twice side by side and label them, just like Newtdog has in Figure 6.3.

- Draw your axis system (Cartesian, polar, or path).

- Draw in the applied forces acting on the left diagram (including weight).

- Replace supports with forces as these also act on the body (example: normal force).

- Draw in mass times acceleration on the right diagram in each direction of motion.

- Assume the positive direction motion when unknown.

6.4   MASS UNITS

For Dynamics the metric system (a.k.a. "System Internationale" or "SI") is much easier to use than the U.S. Customary Units system (formerly known as the "British Gravitational" system) since mass in kilograms is distinctly different than force in Newtons. There can be some confusion since items in the grocery store are measured and sold using kilograms based on readouts of scales, but the layperson is mostly unaware the distinction between mass and weight. Used this way a kilogram is a kilogram, and as long as gravity doesn't vary much, there's no issue.

Figure 6.4: Introducing "Slugs" (© E. Diehl).

When you want the weight force in SI, you multiply kilograms by the standard acceleration of gravity which engineers use as $g = 9.81$ m/s^2 (most engineering texts use this three significant digit value as standard, while you may have used other values in physics classes). So with our FBD/IBD diagrams we use Newtons on the left and Kilograms times acceleration in m/s^2 on the right.

In U.S. Customary Units there is a bit more confusion, especially since there is an on-going disagreement in how to address mass units. In Thermodynamics there are units called "pounds mass" and "pounds force," and they're not treated the same way as is commonly done in Dynamics.

For this course we'll use a unit called the "Slug" for mass, named this because mass resists motion and is therefore "sluggish." Figure 6.4 shows Wormy getting introduced to a slug who is sluggish and can be measured in Slugs. We find the mass of an object by dividing its weight by the engineers' acceleration of gravity $g = 32.2$ ft/s^2 (again, this is the engineering accepted value). So 1 slug $= \frac{1 \text{ lb}}{32.2 \text{ ft/s}^2}$. Note that slugs are based on 32.2 ft/s^2 not 386.4 in/s^2, so it is often best to use feet rather than inches as the base unit for the rest of U.S. units problems to avoid confusion.

For units conversion, always remember "$F = ma$" so when you see the combination of units slug$\frac{\text{ft}}{\text{s}^2}$ (mass times acceleration) it converts into lb (force). This is the same approach as SI units when the combination kg $\frac{\text{m}}{\text{s}^2}$ (mass times acceleration) converts into N (force). Study the summary and Figure 6.5 (with a frictionless block) to help resolve any confusion you might have with the mass units convention used here.

SI:	General Force conversion:	$N = kg \cdot m/s^2$
	Weight:	$(1 \text{ kg}) (9.81 \text{ m/s}^2) = 9.81$ N
U.S. Customary Units:	General Force conversion:	$lb = slug \cdot ft/s^2$
	Weight:	$(1 \text{ slug}) (32.2 \text{ ft/s}^2) = 32.2$ lb.

$$\rightarrow \sum F_x = ma_x \qquad\qquad \rightarrow \sum F_x = ma_x$$

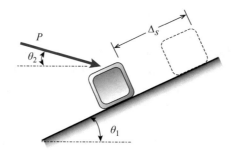

Figure 6.5: FBD/IBD pairs for U.S. and SI units.

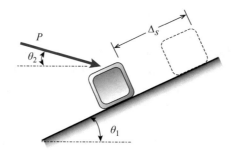

Figure 6.6: Example 6.1.

$$(1 \text{ lb}) = (1 \text{ slug}) \, a_x \qquad\qquad (1 \text{ N}) = (1 \text{ kg}) \, a_x$$

$$a_x = 1 \text{ ft/s}^2 \qquad\qquad a_x = 1 \text{ m/s}^2.$$

Example 6.1

A constant $P = 50$ lb force is applied to a box weighing 25 lb in Figure 6.6, starting from rest, and positioned on a $\theta_1 = 25°$ inclined surface with $\mu_s = 0.25$ and $\mu_k = 0.2$ static and kinetic coefficients of friction, respectively. The force is applied to box $\theta_2 = 15°$ from horizontal. Determine the distance up the slope the box travels (Δs) when it reaches a speed of $v = 3$ ft/s.

The nature of the question implies the box will move, but does it? We should check that the force is sufficient to overcome the static friction. Quite often a problem will tell you to assume it moves, but we should show due diligence and check.

We first draw the FBD/IBD's set in Figure 6.7. Tilting the reference axis to x' and y' will help isolate the directions, making y' a Statics problem to determine the normal force. Note we should be careful to keep track of the angles:

$$\nwarrow \sum F_{y'} = ma_{y'} = 0$$

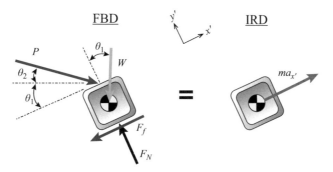

Figure 6.7: **FBD/IBD** of Example 6.1.

$$- P \sin (\theta_1 + \theta_2) - W \cos \theta_1 + F_N = 0$$

$$F_N = (50) \sin (40°) + (25) \cos (25°) = 54.80 \text{ lb.}$$

The static friction force is:

$$F_{f,s} = \mu_s F_N = (0.25) (54.80) = 13.70 \text{ lb.}$$

This is the force that must be overcome before there is motion, not necessarily the force due to friction.

The dynamic friction force is:

$$F_{f,k} = \mu_s F_N = (0.20) (54.80) = 10.96 \text{ lb.}$$

This will be the friction force if the block moves.

If we assume motion up the slope, N2L in the x's direction is:

$$\nearrow \sum F_{x'} = m a_{x'}$$

$$P \cos (\theta_1 + \theta_2) - W \cos \theta_1 - F_f = m a_{x'}.$$

To check if there's motion, we can look at the forces in the x'-direction using the static friction force $(F_{f,s})$ as if we've reached that limit and see if we get a positive force meaning the net force is sufficient to move the box up the incline:

$$(50) \cos (40°) - (25) \sin (25°) - (13.70) = 14.04 \text{ lb.}$$

It's positive, therefore the box will begin to move. We now calculate the acceleration of the box, but use the kinetic friction force instead:

$$(50) \cos (40°) - (25) \sin (25°) - (10.96) = \left(\frac{25}{32.2} \right) a_{x'}$$

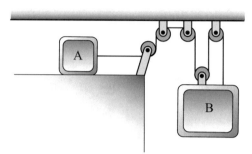

Figure 6.8: **Example** 6.2.

$$a_{x'} = 21.61 \text{ ft/s}^2.$$

We use kinematics to determine the distance traveled to reach $v = 3$ ft/s:

$$v^2 = v_0{}^2 + 2a\Delta s$$

$$\Delta s = \frac{v^2 - v_0{}^2}{2a} = \frac{(3)^2 - 0^2}{2(21.61)} = 0.2083 \quad \text{ft} = 2.499 \text{ in}$$

$$\boxed{\Delta s = 2.50 \text{ in}}.$$

Note: We are treating this box as a particle, that is: its size and shape don't matter to us so all of the forces are assumed to act at one point. If it had shape, it would be a rigid body that could potentially rotate (or possibly tip over) and the normal force on the bottom (and consequently the friction force) wouldn't be constant.

Example 6.2
Two blocks (block A, $m_A = 20$ kg, and block B, $m_A = 35$ kg) connected by ropes and pulleys, as shown in Figure 6.8, are released from rest and begin to move. The coefficient of kinetic friction between block A and the surface is $\mu_k = 0.25$. The pulleys are assumed to be massless and frictionless. Determine the acceleration of each block and the tension in the cable.

This problem is a reminder that Dynamics topics build upon each other, since here we have dependent motion from Class 2. This problem also introduces multiple particles which will require FBD/IBD sets for each. Last, we note "The pulleys are assumed to be massless and frictionless." is included in the problem statement to let us know that the cable tension is the same everywhere along its length. If the pulleys had mass, they would resist rotation (as we'll see in rigid body kinetics) and if they had friction there would be losses and the cable tension would vary in between pulleys.

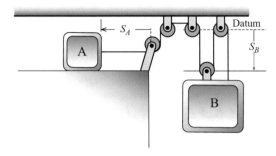

Figure 6.9: Dependent motion in Example 6.2.

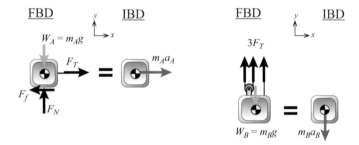

Figure 6.10: FDB/IBD pairs in Example 6.2.

It's often good practice to take care of the kinematics first when we recognize it's needed. Figure 6.9 shows the dependent motion labels to keep track of the lengths of rope segments which we use to find the relation of accelerations between blocks:

$$s_A + 3s_B = constant$$

$$\frac{d}{dt}[s_A + 3s_B] = \frac{d}{dt}[constant]$$

$$v_A + 3v_B = 0$$

$$a_A + 3a_B = 0 \qquad a_{Ax} = |3a_{By}| \; \text{①}.$$

We'll avoid any confusion in signs by taking the absolute value.

The two sets of FBD/IBDs are linked by the cable tension in Figure 6.10.

$$\uparrow \sum F_{Ay} = m_A a_{Ay} = 0$$

$$-m_A g + F_N = 0$$

$$F_N = m_A g = (25)(9.81) = 196.2 \text{ N}$$

$$F_f = \mu_k F_N = (0.25)(196.2) = 49.05 \text{ N}$$

$$\rightarrow \sum F_{Ax} = m_A a_{Ax}$$

$$-F_f + F_T = m_A a_{Ax}$$

$$F_T = m_A a_{Ax} + F_f$$

$$F_T = (20) a_{Ax} + (49.05) \; \text{②} \; .$$

$$\sum F_{By} = m_B a_{By} - m_B g + 3 F_T = -m_B a_{By}$$

$$F_T = \frac{1}{3} \left(m_B g - m_B a_{By} \right) = \frac{1}{3} \left((35)(9.81) - (35) a_{By} \right)$$

$$F_T = (114.5) - (11.67) a_{By} \; \text{③}$$

Set equations ② and ③ equal: $(20) a_{Ax} + (49.05) = (114.5) - (11.67) a_{By}$ ④
Substitute equation ① into equation ④: $(20)(3) a_{By} + (49.05) = (114.5) - (11.67) a_{By}$

$$\boxed{a_{By} = 0.9132 \text{ m/s}^2} \; .$$

We can substitute this result into equations ③ and ④ to find the remaining desired unknowns:

$$\boxed{a_{By} = 0.9132 \text{ m/s}^2 \; \downarrow} \qquad \boxed{a_{Ax} = 2.740 \text{ m/s}^2 \; \rightarrow} \qquad \boxed{F_T = 103.8 \text{ N}(T)} \; .$$

Note that the force doesn't get an arrow here, but instead is labeled with "(T)" to represent tension. This is because the direction is dependent on where you cut the rope and to which side of the cut end you refer. We should be very careful to communicate answers to eliminate possible confusion.

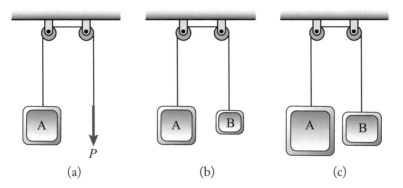

Figure 6.11: Example 6.3.

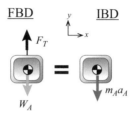

Figure 6.12: FBD/IBD pair of block A in configuration (a) of Example 6.3.

Example 6.3

The three setups shown in Figure 6.11 begin at rest. In setup (a) the force ($P = 50$ lb) is applied on a cable attached to block A ($W_A = 75$ lb). Setup (b) has the same block A and is connected to block B ($W_B = 50$ lb). Setup (c) has larger blocks with the same difference between them ($W_A = 175$ lb and $W_B = 150$ lb). The pulleys are assumed to be massless and frictionless. Determine for each setup the acceleration of block A.

This problem is meant to demonstrate the effect of different masses on the results. It is especially useful to dispel a common misconception that the cable tension will equal the weight. If the blocks are accelerating the cable tension will be either less or more than the weight(s). Generally, an FBD/IBD pair is required for each particle. Remember that if the pulley is massless and frictionless, the cable tension is the same throughout. We'll assume, based on observation, that block A is moving downward. If we were unsure, we'd assume it was upward (the positive direction) and learn it was opposite if the results were negative.

For configuration (a) the only FBD/IBD pair is for block A in Figure 6.12:

$$\uparrow \sum F_{Ay} = m_A a_{Ay}$$

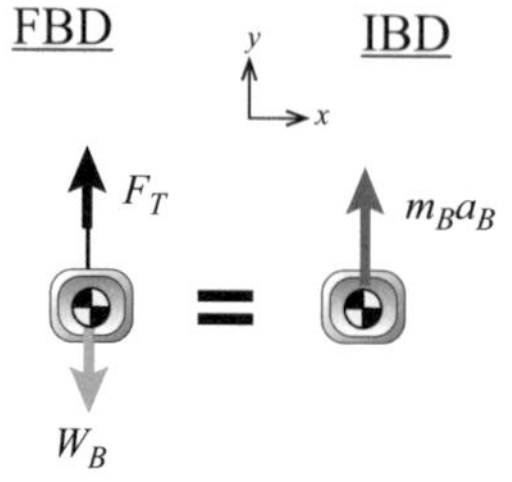

Figure 6.13: FBD/IBD pair of block B in configuration (b) of Example 6.3.

$$F_T - W_A = -m_A a_{Ay} \text{ ①}$$

$$(50) - (75) = -\frac{(75)}{(32.2)} a_{Ay}$$

$$a_{Ay} = 10.73 \text{ ft/s}^2$$

(a) $\boxed{\vec{a}_A = 10.73 \text{ ft/s}^2 \downarrow}$.

For configuration (b) we can use the FBD/IBD of Figure 6.12 and ① from part (a) for block A and include a new FBD/IBD pair for block B in Figure 6.13. We recognize that $a_{Ay} = |a_{By}|$ without having to work through a the dependent motion problem. We use the absolute value because the FBD/IBD's address the directions of the accelerations:

$$\uparrow \sum F_{By} = m_B a_{By}$$

$$F_T - W_B = m_B a_{By}$$

$$F_T = m_B a_{By} + W_B \text{ ②}.$$

Combining ① and ② and replace $a_{Ay} = |a_{By}|$

$$m_B a_{Ay} - W_B - W_A = -m_A a_{Ay} \text{ ③}$$

$$\frac{(50)}{(32.2)} a_{Ay} + (50) - (75) = -\frac{(75)}{(32.2)} a_{Ay}$$

$$a_{Ay} = 6.440 \text{ ft/s}^2$$

(b) $\boxed{\vec{\mathbf{a}}_A = 6.440 \text{ ft/s}^2 \downarrow}$.

We find that using a 50 lb mass on the other end of the cable rather than a 50 lb force causes lower acceleration. Why? Because now there is more mass overall to accelerate, so the motion is more "sluggish." We weren't asked, but let's see what the cable tension is now using equation ②:

$$F_T = m_B a_{By} + W_B = \frac{(50)}{(32.2)} (6.440) + (50) = 60.00 \text{ lb}.$$

The tension has increased but is still less than the weight of block A, which makes sense that it is still moving downward if we look at the first FBD/IBD.

For configuration (c) we don't need new FBD/IBD's as only the weights have changed, so we can use equation ③ again. Note the difference in weights are the same, so we might conclude (wrongly!) before beginning that the acceleration should be the same:

$$m_B a_{Ay} - W_B - W_A = -m_A a_{Ay}$$

$$\frac{(150)}{(32.2)} a_{Ay} + (150) - (175) = -\frac{(175)}{(32.2)} a_{Ay}$$

$$a_{Ay} = 2.477 \text{ ft/s}^2$$

(c) $\boxed{\vec{\mathbf{a}}_A = 2.477 \text{ ft/s}^2 \downarrow}$.

Once again, with increased mass we have less acceleration. Check the tension in this configuration using equation ③:

$$F_T = m_B a_{By} + W_B = \frac{(150)}{(32.2)} (2.477) + (150) = 161.5 \text{ lb}.$$

This also makes sense since it is less than the 175 lb of block A which is moving downward.

This problem will be revisited in Work-Energy and Impulse-Momentum to demonstrate that we can sometimes solve kinetics problems in multiple ways in conjunction with kinematics. We will also revisit the same problem in rigid body motion to see the effect when pulleys have mass.

Example 6.4

A basket of apples with mass of 15 kg is on the back of the wagon in Figure 6.14 with a sloped surface of $\theta = 15°$. The center of the basket is $\Delta s = 2$ m from the edge. The wagon begins from rest and accelerates. The coefficients of friction between the basket and the surface are $\mu_s = 0.35$

Figure 6.14: Newtdog on a wagon in Example 6.4 (© E. Diehl).

Figure 6.15: FBD/IBD of Example 6.4 Part 1 (© E. Diehl).

and $\mu_k = 0.30$. Determine the maximum acceleration the wagon can have without the basket sliding off and the acceleration required for the basket to slide off in 1 s.

For the first part of the question the wagon and basket accelerate at the same rate until the motion overcomes static friction. The wagon is redrawn in Figure 6.15 to define the geometry. The FBD/IBD pair in Figure 6.16 shows the basket isolated and has the horizontal acceleration of the wagon broken into components of a coordinate system aligned with the sloped surface:

$$\nearrow \sum F_{y'} = ma_{y'} \qquad -mg\cos\theta + F_N = -ma_W\sin\theta \qquad F_N = m\left(-a_W\sin\theta + g\cos\theta\right)$$

$$F_f = \mu_s F_N$$

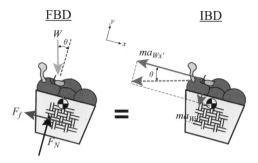

Figure 6.16: Arrangement of Example 6.4 (© E. Diehl).

We use static friction here because of we're being asked for the acceleration when static friction is overcome:

$$\searrow \sum F_{x'} = ma_{x'} \qquad mg \sin \theta - \underbrace{\mu_s m \left(-a_W \sin \theta + g \cos \theta\right)}_{F_f} = -ma_W \cos \theta.$$

Note that the mass cancels out which means the amount of apples doesn't affect the results. This isn't necessarily intuitive, but we will see this is often the case in other kinetics situations:

$$g \sin \theta + \mu_s a_W \sin \theta - \mu_s g \cos \theta = -ma_W \cos \theta$$

$$a_W = g \frac{(\mu_s \cos \theta - \sin \theta)}{(\cos \theta + \mu_s \sin \theta)} = (9.81) \frac{((0.35) \cos 15° - \sin 15°)}{(\cos 15° + (0.35) \sin 15°)} = 0.7359 \text{ m/s}^2$$

$$\boxed{\vec{a}_W = 0.736 \text{ m/s}^2 \leftarrow}.$$

For the second part of the question, the wagon and basket accelerate at different rates *and directions*, so we need to consider the relative motion. The relative acceleration (basket with respect to wagon, $a_{B/W}$) can be found from kinematics:

$$x = x_o + v_o t + \frac{1}{2} a_{B/W} t^2 \quad x_o = 0, \ v_o = 0, \quad 2 = \frac{1}{2} a_{B/W} (1)^2 \quad a_{B/W} = 1.000 \text{ m/s}^2.$$

The actual acceleration of the basket is a combination of the acceleration of the wagon and this relative acceleration:

$$\vec{a}_B = \vec{a}_W + \vec{a}_{B/W}.$$

Note that it will be more convenient to apply both the wagon acceleration and the relative acceleration to the IBD of the basket in Figure 6.17:

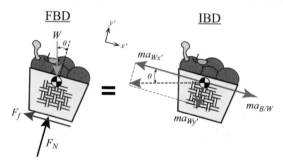

Figure 6.17: FBD/IBD pair of Example 6.4 Part 2 (© E. Diehl).

$$\nearrow \sum F_{y'} = ma_{y'} \quad -mg\cos\theta + F_N = -ma_W\sin\theta \quad F_N = m\left(-a_W\sin\theta + g\cos\theta\right)$$

$$F_f = \mu_k F_N$$

$$\searrow \sum F_{x'} = ma_{x'} \quad mg\sin\theta - \underbrace{\mu_k m\left(-a_W\sin\theta + g\cos\theta\right)}_{F_f} = -ma_W\cos\theta + ma_{B/W}.$$

Again, the basket mass cancels out:

$$g\sin\theta + \mu_k a_W\sin\theta - \mu_k g\cos\theta = -ma_W\cos\theta + a_{B/W}$$

$$a_W = \frac{\mu_k g\cos\theta - g\sin\theta + a_{B/W}}{\mu_k\sin\theta + m\cos\theta}$$

$$a_W = \frac{g\left(\mu_k\cos\theta - \sin\theta\right) + a_{B/W}}{\left(\cos\theta + \mu_k\sin\theta\right)} = \frac{(9.81)\left((0.30)\cos15° - \sin15°\right) + (1.000)}{\left(\cos15° + (0.30)\sin15°\right)} = 1.249\,\text{m/s}^2$$

$$\boxed{\vec{a}_W = 1.25\,\text{m/s}^2 \leftarrow}.$$

This is the acceleration of the wagon that will result in a relative acceleration of the basket down the slope.

The "trick" of using both the wagon acceleration (broken into the rotated component directions) and the relative acceleration on the IBD probably isn't obvious to most of us. Then how do we know to do it? We just have to work through as much of the problem as we can until we get stuck and then try some things we know until the "ah-ha" moment hits us. We will see many more "trick" problems, but a key solution strategy is to work through as much as we can, applying what we know and persevere by trying various things we know about the situation until we find the right approach. With practice this becomes less daunting, especially as your confidence grows and your ability to recognize when to apply concepts improves with the experience.

Book 1 - Class 7

https://www.youtube.com/watch?v=UjTBAvWGfuM

C L A S S　7

Newton's Second Law (N2L) in Non-Rectangular Coordinates

B.L.U.F. (Bottom Line Up Front)

- N2L in path coordinates:　$\sum F_t = ma_t = m\frac{dv}{dt}$ and $\sum F_n = ma_n = m\frac{v^2}{\rho}$.

- N2L in polar coordinates:　$\sum F_r = ma_r = m\left(\ddot{r} - r\dot{\theta}^2\right)$ and $\sum F_\theta = ma_\theta = m\left(r\ddot{\theta} + 2\dot{r}\dot{\theta}\right)$.

7.1　N2L IN PATH COORDINATES

Path coordinates are especially useful when analyzing the dynamic forces of objects moving about a curve or turning. Most often there is some curved surface the object is moving over or on/around such as a dip or turn in the road. We are all familiar with the sensation of being pushed outward in a car during a fast or sharp turn, and this can cause a bit of confusion as it "feels" like there's an outward force or that we're being accelerated outward. In this scenario we are actually being accelerated inward by the car and we're feeling the reaction (Newton's Third Law) of the seat (or the car door) pushing against us toward the center of curvature, this so-called "centrifugal force." We are being transported by the car and would like to continue tangent to the path but the car is moving us into the curve, thus pushing (and accelerating) us toward the curve center. The FBD/IBDs help to resolve this possible confusion. Keep in mind that kinetics describes the cause of the motion (while kinematics only describes the motion) so we're investigating what forces are on an object. We used path coordinates in kinematics to describe the acceleration. Now there must be a reason it was moving in that path, so we apply kinetics in path coordinates to figure out why.

7.2　PATH COORDINATES FBD/IBD

Setting up the FBD/IBDs in path coordinates follows the same methodology of Cartesian co-ordinates except the forces and accelerations are resolved into the components aligned with the

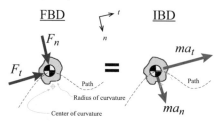

Figure 7.1: Path coordinates FBD/IBD.

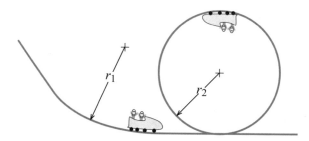

Figure 7.2: Roller coaster of Example 7.1.

normal and tangential directions. Figure 7.1 demonstrates a generic path coordinates FBD/IBD pair. The path has been super-imposed here but is not necessary in your diagrams:

$$\searrow \sum F_t = ma_t = m\frac{dv}{dt} \quad \text{and} \quad \nearrow \sum F_n = ma_n = m\frac{v^2}{\rho}.$$

Remember, the acceleration component in the tangential direction is due to the change in speed $\left(a_t = \frac{dv}{dt}\right)$ and in the normal direction is due to the change in direction $\left(a_n = \frac{v^2}{\rho}\right)$. The force corresponding with the normal direction is the source of the centrifugal force, although, as mentioned, we're actually feeling the reaction which is in the opposite direction.

There is often some type of constraint causing the curved motion. A train or roller coaster following a track, a car traveling on a hill, or a ball tied to a string are examples. A space craft might use multiple thrusters to achieve a curved path. In any of these cases, the FBD/IBD procedure is to break the forces and accelerations into these components. Roller coasters are fun examples of using N2L in path coordinates.

Example 7.1
In the coaster shown in Figure 7.2, the car reaches 35 mph at the bottom of a curved drop (with radius $r_1 = 60$ ft) at just before a short, flat portion. The car then travels into a $r_2 = 25$ ft loop-d-loop. For a 50 lb kid, how much does he perceive he weighs at the bottom of the drop, and how fast does the car need to travel around the loop so he doesn't fall out without a harness?

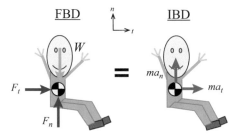

FBD   IBD

Figure 7.3: FBD/IBD of Example 7.2 at bottom of drop.

We convert the speed into base units first

$$\frac{(35 \text{ mph}) (5280 \text{ ft/mi})}{(3600 \text{ s/hr})} = 51.33 \text{ ft/s}.$$

The kid's perceived weigh is the force of the seat pushing against him. Note that the problem states the car has reached the bottom of the curve "just before" the flat portion, so it is still in the curve but nearly horizontal. We draw the FBD/IBD pair of the kid alone to do the analysis in Figure 7.3. We've drawn forces in the positive tangential direction (the roller coaster *could* be accelerating or decelerating, we don't know) but these won't be needed:

$$\uparrow \sum F_n = ma_n$$

$$F_n - W = m\frac{v^2}{\rho}$$

$$F_n = W + m\frac{v^2}{\rho} = (50) + \left(\frac{50}{32.2}\right)\frac{(51.33)^2}{(60)} = 118.2 \text{ lb}$$

$$\boxed{F_n = 118 \text{ lb}}.$$

To find the minimum speed to go around the loop-d-loop we need to analyze when the kid is completely inverted at the top. We draw a new FBD/IBD pair in Figure 7.4. The minimum speed will be when the normal force equals zero:

$$\downarrow \sum F_n = ma_n$$

$$F_n + W = m\frac{v^2}{\rho}$$

$$(0) + (50) = \left(\frac{50}{32.2}\right)\frac{v^2}{(25)}$$

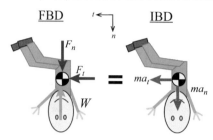

Figure 7.4: FBD/IBD of Example 7.2 at top of loop.

Figure 7.5: Newtdog riding a coaster (© E. Diehl).

Note that the mass would cancel out here:

$$v = 40.12 \text{ ft/s} = 27.36 \text{ mph}$$

$$\boxed{v_{\min} = 27.4 \text{ mph}} .$$

You could make a pretty good case that "centrifugal force" is what makes roller-coasters fun. Getting lifted off our seats (like Newtdog in Figure 7.5) when the curve is convex and pushed into our seats when it's concave is fun part. But we engineers know that this isn't a real force, but instead we're feeling the reaction forces pushing against us making it feel like we're being lifted and pushed. We have to be careful to not mis-use our intuition sometimes. The FBD/IBD pair can really help to resolve our false impressions on the directions forces are acting.

Figure 7.6: Newtdog swings a pendulum (© E. Diehl).

Example 7.2

Newtdog has tied Wormy's large apple ($W = 0.5$ lb) to a 2 ft string to make a pendulum as shown in Figure 7.6. He starts it in motion, and the maximum speed (when the apple is at the bottom of the arc) is 1 ft/s. What is the maximum tension in the string and the maximum angle it will rise (measured from the vertical axis)?

The first part of the question is relatively easy since the maximum tension will occur when the normal acceleration is maximum: at the bottom of the swing. We draw the FBD/IBD pair of the apple at the bottom of the swing (Figure 7.7). Notably, there cannot be a tangential force in this position so we conclude there is no acceleration as well. We already knew this because of kinematics: acceleration is zero when velocity is a maximum:

$$\uparrow \sum F_n = ma_n$$

$$F_T - W = m\frac{v^2}{\rho}$$

$$F_T = W + m\frac{v^2}{\rho} = (0.5) + \left(\frac{0.5}{32.2}\right)\frac{(1)^2}{(2)} = 0.5311 \text{ lb}$$

$$\boxed{F_T = 0.531 \text{ lb}}.$$

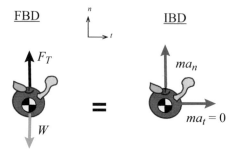

Figure 7.7: FBD/IBD of Example 7.2 at bottom of pendulum swing.

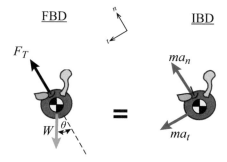

Figure 7.8: FBD/IBD of Example 7.2 at top of pendulum swing (© E. Diehl).

The second part of the question is considerably more difficult. We'll need to use some kinematics and even some integration to get the maximum angle. We begin the N2L method with an FBD/IBD pair drawn at an arbitrary angle in Figure 7.8. We note that the tension of the string isn't provided, so we'll avoid it by using only the tangential direction:

$$\swarrow \sum F_t = ma_t$$
$$W \sin \theta = ma_t$$
$$mg \sin \theta = ma_t.$$

It is interesting that mass cancels, leaving a relationship between tangential acceleration as a function of angle:

$$a_t = g \sin \theta.$$

We don't know the acceleration at the point but we can apply some kinematics to find how it relates to angle. Acceleration along an arc is the tangential acceleration and can be written as $a_t = v\frac{dv}{ds}$. As shown in Figure 7.9, we know the arc length of a short segment is $ds = r\, d\theta$. If we replace this into the tangential acceleration we get $a_t = \frac{v}{r}\frac{dv}{d\theta}$. Rearranging we get $v\, dv = a_t r\, d\theta$.

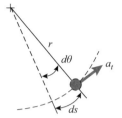

Figure 7.9: Tangential acceleration in Example 7.2.

Since we have a relationship of the tangential acceleration as a function of angle, $a_t(\theta)$, we can use these together to integrate between maximum velocity to zero velocity, which is the location of the maximum angle:

$$\int_{v_{\max}}^{0} v \, dv = \int_{0}^{\theta_{\max}} a_t r d\theta = \int_{0}^{\theta_{\max}} (g \sin \theta) \, r d\theta$$

$$0 - \frac{1}{2} v_{\max}^2 = -gr \cos \theta_{\max} + gr$$

$$\theta_{\max} = \cos^{-1} \left[\frac{\frac{1}{2} v_{\max}^2 - gr}{gr} \right] = \cos^{-1} \left[\frac{\frac{1}{2}(1)^2 - (32.2)(2)}{(32.2)(2)} \right] = 71.44°$$

$$\boxed{\theta_{\max} = 71.4°} \ .$$

In the Work-Energy topic we'll revisit this problem and see it's a much easier approach that returns the same answer.

7.3   N2L IN POLAR COORDINATES

Recall from kinematics that acceleration in polar coordinates is:

$$\vec{\mathbf{a}} = \underbrace{\left(\ddot{r} - r \dot{\theta}^2 \right)}_{a_r} \hat{\mathbf{e}}_r + \underbrace{\left(r \ddot{\theta} + 2 \dot{r} \dot{\theta} \right)}_{a_\theta} \hat{\mathbf{e}}_\theta.$$

When applying Newton's Second Law we take the force in the radial direction and equate it with mass times the acceleration in the radial direction and do the same for the transverse direction. Figure 7.10 shows a typical FBD/IBD pair in polar coordinates. The x–y axes are shown here for reference but aren't necessary each time we draw an FBD/IBD pair:

$$\nearrow \sum F_r = ma_r = m \left(\ddot{r} - r \dot{\theta}^2 \right) \quad \text{and} \quad \nwarrow \sum F_\theta = ma_\theta = m \left(r \ddot{\theta} + 2 \dot{r} \dot{\theta} \right).$$

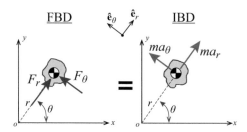

Figure 7.10: FBD/IBD in polar coordinates.

Figure 7.11: Newtdog considering polar coordinates (© E. Diehl).

Figure 7.12: Newtdog with Wormy on a fishing line, repeat of Figure 4.2 (© E. Diehl).

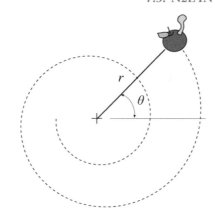

Figure 7.13: Sketch of path of Apple in Example 7.3 (© E. Diehl).

Remember that any coordinate system can be used when solving problems, but some situations are better suited to certain coordinate systems. For polar coordinates (Figure 7.2) that usually means there is a reference point about which a mass is spinning and possibly moving toward or away from. When approaching a new N2L problem, pay close attention to which coordinate system is most appropriate for the circumstances just like Newtdog is in Figure 7.11. By choosing the right coordinate system you can often save yourself some mathematical steps. There might also be situations where problems can only be solved using a particular coordinate system.

Example 7.3
Newtdog is spinning on a stool while holding a 6 ft fishing rod with an apple ($W = 0.5$ lb) tied to the end of the line as shown in Figure 7.12. He is spinning at a rate of $\dot{\theta} = 20$ rpm counter-clockwise but is slowing down. At the instant shown the line is 3 ft long from the tip of the rod, and he is allowing the line to pay out at $\dot{r} = 1$ ft/s, increasing at a constant $\ddot{r} = 0.5$ ft/s^2. Assume the rod and line are all in the horizontal plane and there is no air resistance on the apple. Determine the angular acceleration or deceleration (if any) of the spinning stool and the tension in the fishing line.

It's good to try to imagine what the motion looks like. If the fishing line is payed out, the path will be like a spiral when looked at from above as sketched in Figure 7.13.

Just as we did in polar coordinates with kinematics, we'll try to find/summarize the *variables of interest*: $r, \dot{r}, \ddot{r}, \theta, \dot{\theta}, \ddot{\theta}$:

$$r = 3 + 6 = 9 \text{ ft}, \quad \dot{r} = 1 \text{ ft/s}, \quad \ddot{r} = 0.5 \text{ ft/s}^2$$

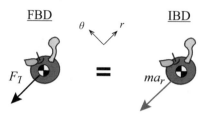

Figure 7.14: FBD/IBD of Apple in Example 7.3 (© E. Diehl).

$$\theta = \text{unkown}, \quad \dot{\theta} = \frac{(20 \text{ rpm})}{(60 \text{ s/min})}(2\pi \text{ rad/rev}) = 2.094 \text{ rad/s}, \quad \ddot{\theta} = \text{unknown}.$$

From here we know that we'll need to somehow figure out what θ and $\ddot{\theta}$ are. Let's draw an FBD/IBD pair and see if that can help (Figure 7.5).

A few things become clear from the FBD/IBD pair we draw in Figure 7.14. The force of the fishing line on the apple must be in the negative r direction since you "can't push a rope." We also know there is no force in the transverse (θ) direction because the problem states there is no air resistance and the fishing line can only be in tension which is in the radial direction. Last, we note that the angle doesn't matter:

$$\nwarrow \sum F_\theta = ma_\theta$$

$$0 = ma_\theta = m\left(r\ddot{\theta} + 2\dot{r}\dot{\theta}\right).$$

Similar to Example 7.2, the mass cancels here:

$$r\ddot{\theta} + 2\dot{r}\dot{\theta} = 0$$

$$\ddot{\theta} = -\frac{2\dot{r}\dot{\theta}}{r} = -\frac{2(1)(2.094)}{(9)} = -0.4654 \text{ rad/s}^2$$

$$\boxed{\ddot{\theta} = 0.465 \text{ rad/s}^2 \text{ CW}}.$$

The spin rate is therefore decreasing in order to have no transverse ($\hat{e}_\theta$) component of force or acceleration. The tension in the fishing line can be found from the radial direction on the FBD/IBD:

$$\nearrow \sum F_r = ma_r$$

$$-F_T = -ma_r = -m\left(\ddot{r} - r\dot{\theta}^2\right)$$

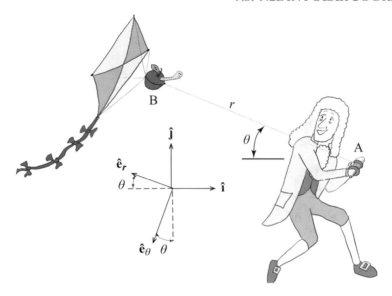

Figure 7.15: Newtdog flying a kite with Wormy (© E. Diehl).

$$F_T = \left(\frac{0.5}{32.2}\right)\left((0.5) - (9)(2.094)^2\right)$$

$$F_T = 0.6050 \text{ lb.}$$

It might be possible to achieve this result using path coordinates, but it would require us to write the equation for the spiral shown in Figure 7.13 since the normal acceleration is NOT along the fishing line since the center of curvature is not at the origin. This would also require a coordinate transformation.

Example 7.4
Newtdog is flying Wormy's apple ($m = 0.2$ kg) on a kite in Figure 7.15. He accelerates to the right at 4.5 m/s^2 and is going 1 m/s at the instant shown. The kite string is at $\theta = 25°$ from horizontal, increasing at 0.25 rad/s and 0.1 rad/s^2. The kite string has a tension of 40 N, is 5 m long, and is unspooling at 0.5 m/s, although the rate of unspooling is reducing at 1 m/s^2. What forces are being exerted by the kit on the apple? Provide the answer in Cartesian coordinates.

There is a lot of information provided so let's summarize what we know. We'll call the location of the spool point A and the apple point B. We recognize this uses polar coordinates,

Table 7.1: Transformation matrix

	$\hat{\mathbf{i}}$	$\hat{\mathbf{j}}$
$\hat{\mathbf{e}}_r$	$-\cos\theta$	$\sin\theta$
$\hat{\mathbf{e}}_\theta$	$-\sin\theta$	$-\cos\theta$

so we summarize the V.o.I.:

$$r = 5 \text{ m} \qquad\qquad \theta = 25°$$
$$\dot{r} = 0.5 \text{ m/s} \qquad\qquad \dot{\theta} = -0.25 \text{ rad/s}$$
$$\ddot{r} = -1 \text{ m/s}^2 \qquad\qquad \ddot{\theta} = -0.1 \text{ rad/s}^2.$$

The acceleration described using polar coordinates is in reference to the spool which is also moving. So the acceleration of B is with respect to A:

$$\vec{a}_{B/A} = \left(\ddot{r} - r\dot{\theta}^2\right)\hat{\mathbf{e}}_r + \left(r\ddot{\theta} + 2\dot{r}\dot{\theta}\right)\hat{\mathbf{e}}_\theta$$
$$= \left((-1) - (5)(-0.25)^2\right)\hat{\mathbf{e}}_r + ((5)(-0.1) + 2(0.5)(-0.25))\hat{\mathbf{e}}_\theta \text{ m/s}^2$$

$$\vec{a}_{B/A} = \underbrace{(-1.313)}_{a_r}\hat{\mathbf{e}}_r + \underbrace{(-0.75)}_{a_\theta}\hat{\mathbf{e}}_\theta \text{ m/s}^2.$$

In order to add the acceleration of the spool, that too should be in polar coordinates. We construct a transformation matrix to help (Table 7.1):

$$\vec{a}_A = (4.5)\hat{\mathbf{i}} = (4.5)(-\cos\theta)\hat{\mathbf{e}}_r + (-4.5)(\sin\theta)\hat{\mathbf{e}}_\theta$$

$$\vec{a}_A = (4.5)\left(-\cos(25°)\right)\hat{\mathbf{e}}_r + (4.5)\left(-\sin(25°)\right)\hat{\mathbf{e}}_\theta$$

$$\vec{a}_A = (-4.078)\hat{\mathbf{e}}_r + (-1.902)\hat{\mathbf{e}}_\theta \text{ m/s}^2.$$

The total acceleration of the apple is therefore:

$$\vec{a}_B = \vec{a}_A + \vec{a}_{B/A}$$
$$= (-4.078 - 1.313)\hat{\mathbf{e}}_r + (-1.902 - 0.75)\hat{\mathbf{e}}_\theta = \underbrace{(-5.391)}_{a_r}\hat{\mathbf{e}}_r + \underbrace{(-2.652)}_{a_\theta}\hat{\mathbf{e}}_\theta \text{ m/s}^2.$$

Accelerations in the FBD/IBD in Figure 7.16 are drawn in the positive directions and given signs as appropriate in the equations:

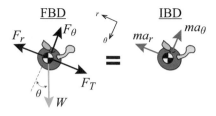

Figure 7.16: FBD/IBD for Example 7.4 (© E. Diehl).

$$\swarrow \sum F_\theta = ma_\theta$$

$$-F_\theta + W \cos \theta = ma_\theta$$

$$F_\theta = (0.2)(9.81) \cos \left(25°\right) - (0.2)(-2.652) = 2.309 \text{ N}$$

$$\nwarrow \sum F_r = ma_r$$

$$F_r - W \sin \theta - F_T = ma_r$$

$$F_r = (0.2)(9.81) \sin \left(30°\right) + (40) + (0.2)(-2.563)$$

$$F_r = 40.32 \text{ N}$$

$$\overrightarrow{\mathbf{F}} = (40.32)\,\hat{\mathbf{e}}_r + (2.309)\,\hat{\mathbf{e}}_\theta \ \text{N}.$$

Transformed into Cartesian Coordinates:

$$\overrightarrow{\mathbf{F}} = \left[-(40.32) \cos \left(25°\right) - (2.309) \sin \left(25°\right)\right]\hat{\imath} \\ + \left[(40.32) \sin \left(25°\right) - (2.309) \cos \left(25°\right)\right]\hat{\jmath} \ \text{N}.$$

$$\overrightarrow{\mathbf{F}} = (-37.52)\,\hat{\imath} + (14.95)\,\hat{\jmath} \ \text{N}$$

Magnitude:

$$|\overrightarrow{\mathbf{F}}| = \sqrt{(-37.52)^2 + (14.95)^2} = 40.38 \text{ N}.$$

Figure 7.17: Newtdog rides a banked curve (© E. Diehl).

Direction:

$$\tan^{-1}\left(\frac{(14.95)}{(37.52)}\right) = 21.73° \text{ CW w.r.t. } x\text{-axis}$$

$$\boxed{\vec{\mathbf{F}} = 40.4 \text{ N} \searrow 21.7°} \ .$$

7.4   3-D PATH COORDINATES

Although we've said we're sticking to two-dimensional motion, an exception here is made to analyze problems when the curvature causing the normal acceleration isn't in the same plane as the other force(s) that interest us. A common example is the analysis of a vehicle turning around a curve on a sloped embankment, such as a race track.

As seen in Figure 7.17, Newtdog is bicycling around a banked curve. If we want to analyze the friction of the road on the wheels, we need to first look down onto the curve to understand there is an acceleration in the normal direction and then look at a plane in the vertical and normal direction to find the FBD/IBD that will allow us to resolve these forces.

Note that we are still treating this as a particle, that is, as if all of the forces act on one point even though the cartoon might lead you to start thinking about considering the moments. We can pretend he's close enough to the ground to not matter. Also note that the friction force is drawn in both directions. That's because if he's traveling slowly, he'll begin to slide down the embankment without sufficient friction and if he's traveling fast enough he'll begin to slide up the embankment without sufficient friction. There is a speed at which he requires no friction, which is called the track's *rated speed*.

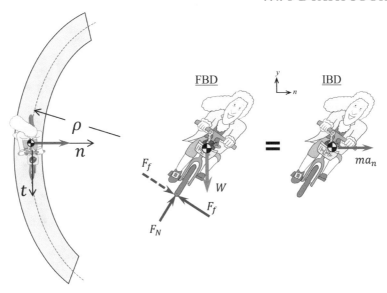

Figure 7.18: FBD/IBD of Newtdog riding on a banked curve on a track (© E. Diehl).

In Figure 7.18 we see the view from above uses the path coordinates normal and tangential components. Instead of using this view, we'll find more meaningful information from the front view which is the normal and vertical components. We've used "y" to designate the vertical axis. We'll see in Example 7.5 that it's advantageous to rotate these axes to align with the banked surface.

Example 7.5

The bicycle racing track shown in the previous figure (Figure 7.18) has a turn with radius of curvature $\rho = 45$ m. The typical racing speed is 40 km/hr. What banking angle is required so no friction is required to keep a bicycle at this speed from slipping outward or inward (making this the "rated speed")? What is the minimum speed a cyclist would need in order to not slip inward on the curve on this track if the coefficient of static friction is $\mu_s = 0.20$ on a wet day?

First, we'll convert the speed into base units:

$$\frac{(40 \text{ km/hr}) (1000 \text{ m/km})}{(3600 \text{ s/hr})} = 11.11 \text{ m/s}.$$

We'll draw the FBD/IBD pair in Figure 7.19 and align the axes with the slope by rotating it at an arbitrary angle θ and designating the new directions as x' and y'. This is for convenience since we can write just one equation in the x' direction to solve this rather than two equations with F_N in them.

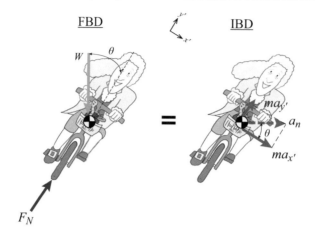

Figure 7.19: FBD/IBD of Example 7.5 without friction (© E. Diehl).

The acceleration in the normal direction is

$$a_n = \frac{v^2}{\rho} = \frac{(11.11)^2}{(45)} = 2.744 \text{ m/s}^2$$

$$\searrow \sum F_{x'} = ma_{x'}$$
$$W \sin \theta = ma_n \cos \theta$$
$$mg \sin \theta = ma_n \cos \theta.$$

Note here that mass cancels out, which means the angle for this is independent of mass. This is one of those instances where our intuition might not coincide with the actual dynamics:

$$\frac{\sin \theta}{\cos \theta} = \tan \theta = \frac{a_n}{g}$$

$$\theta = \tan^{-1}\left(\frac{a_n}{g}\right) = \tan^{-1}\left(\frac{(2.744)}{(9.81)}\right) = 15.62° \qquad \boxed{\theta = 15.6°}.$$

If a cyclist travels less than 40 km/hr they might slip toward the center of the curve except that friction acts on the side of the tires. We re-draw a new FBD/IBD set in Figure 7.20 to show the friction force in this direction:

$$\nearrow \sum F_{y'} = ma_{y'}$$
$$F_N - W \cos \theta = ma_n \sin \theta$$
$$F_N = m (g \cos \theta + a_n \sin \theta)$$

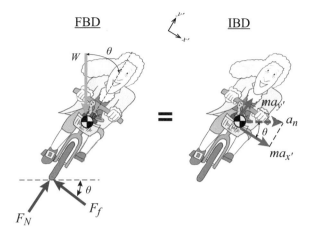

Figure 7.20: FBD/IBD of Example 7.5 with friction to prevent sliding down bank (© E. Diehl).

$$F_f = \mu_s F_N = \mu_s m \left(g \cos \theta + a_n \sin \theta \right)$$

$$\searrow \sum F_{x'} = ma_{x'}$$
$$-F_f + W \sin \theta = ma_n \cos \theta.$$

We replace the friction force and note that once again we can cancel out mass:

$$-\mu_s m \left(g \cos \theta + a_n \sin \theta \right) + mg \sin \theta = ma_n \cos \theta$$

$$a_n = \frac{g \left(\sin \theta - \mu_s \cos \theta \right)}{\cos \theta + \mu_s \sin \theta}$$

$$a_n = \frac{(9.81) \left(\sin (15.62°) - (0.20) \cos (15.62°) \right)}{\cos (15.62°) + (0.20) \sin (15.62°)} = 0.7379 \text{ m/s}^2.$$

From $a_n = \frac{v^2}{\rho}$ we find $v = \sqrt{a_n \rho} = \sqrt{(0.7379)(45)} = 5.763$ m/s

$$v = \frac{(5.763 \text{ m/s}) (3600 \text{ s/hr})}{(1000 \text{ m/km})} = 20.745 \text{ km/hr}$$

$$\boxed{v = 20.7 \text{ km/hr}}.$$

Figure 7.21: Newtdog plays tetherball (© E. Diehl).

Example 7.6
In Figure 7.21, Newtdog is playing tetherball with an old rope that has a breaking strength of 10 lb. The tetherball weighs 1 lb, and the rope is 6 ft long. How fast does the ball have to travel to break the rope?

A sketch (Figure 7.22) helps us to see the movement is like a cone shape. If we looked down on it we see a circular path where path coordinates can be used. But if we used that view, the weight of the ball wouldn't be considered. So instead we draw an FBD/IBD pair of the tetherball in Figure 7.23 from a side view at an arbitrary angle:

$$\uparrow \sum F_y = ma_y = 0$$

$$-W + F_T \sin \theta = 0$$

$$\sin \theta = \frac{W}{F_T}$$

$$\theta = \sin^{-1}\left(\frac{W}{F_T}\right) = \sin^{-1}\left(\frac{(1)}{(10)}\right) = 5.739°.$$

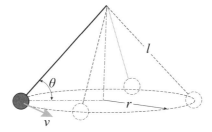

Figure 7.22: Tetherball details in Example 7.6.

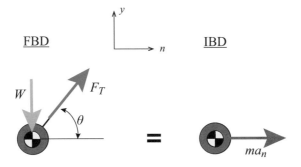

Figure 7.23: FBD/IBD of tetherball in Example 7.6.

We can use this angle to find the radius of the circle the ball travels. $r = l \cos \theta = (6) \cos(5.739°) = 5.970$ ft:

$$\rightarrow \sum F_n = ma_n$$

$$F_T \cos \theta = m \frac{v^2}{\rho}.$$

$$(10) \cos(5.739°) = \left(\frac{1}{32.2} \right) \frac{v^2}{(5.970)}$$

$$v = 43.73 \text{ ft/s} = 29.82 \text{ mph}$$

$$\boxed{v = 29.8 \text{ mph}}.$$

This problem didn't provide the angle of the tetherball rope, and for many students this can cause confusion. It's when you draw the FBD/IBD pair and look the vertical direction that an "ah-ha" moment would likely occur. The first step though is to draw the FBD/IBD and look for clues. The importance of Free Body Diagrams in engineering cannot be overstated.

Book 1 - Class 8

https://www.youtube.com/watch?v=6pkngaXc8Yc

CLASS 8

Work-Energy Method and the Conservation of Energy (Part 1)

B.L.U.F. (Bottom Line Up Front)

- Work-Energy: $KE_1 + PE_1 + U_{1\rightarrow 2} = KE_2 + PE_2$.

- Work: $U_{1\rightarrow 2} = \int_1^2 \overline{\mathbf{F}} \cdot d\overline{\mathbf{r}}$.

- Kinetic Energy: $KE = \frac{1}{2}mv^2$.

- Gravitational Potential Energy: $PE_g = mgy = Wy$.

- Spring Potential Energy: $PE_{sp} = \frac{1}{2}k\delta^2$.

- Conservation of Energy: $KE_1 + PE_1 = KE_2 + PE_2$.

8.1   WORK-ENERGY EQUATION

One of the most famous anecdotes in the history of science tells of young Sir Isaac Newton being struck on the head while sitting beneath an apple tree. This accident supposedly lead him to "discover gravity." Gravity, of course, was already known to exist, but the event actually did occur and lead Newton to ponder the nature of gravity.

The cartoon depiction in Figure 8.1 of this historical event highlights the essentials of the topics covered in this class: changing forms of energy. In this case the apple starts off with potential energy (PE) which is converted into kinetic energy (KE) as it drops toward Newtdog's head.

The complete equation we'll use is presented below, and the individual parts are described in the following sections:

$$\boxed{KE_1 + PE_1 + U_{1\rightarrow 2} = KE_2 + PE_2} \,,$$

Figure 8.1: Newtdog discovers gravity as Potential Energy (PE) is converted into Kinetic Energy (KE) (©E. Diehl).

where

$KE_1 =$ Kinetic Energy at state 1

$PE_1 =$ Potential Energy at state 1

$U_{1\rightarrow 2} =$ External Work acting on system between states 1 and 2

$KE_2 =$ Kinetic Energy at state 2

$PE_2 =$ Potential Energy at state 2

The Work-Energy equation might seem familiar to students who have taken or are taking Fluid Mechanics or Thermodynamics. You can think of this as the solid mechanics form of energy accounting. It's good to use the following grouping to organize how you think of energy accounting:

$$\underbrace{KE_1 + PE_1 +}_{\text{Initial Energy}} \quad \underbrace{U_{1\rightarrow 2}}_{\substack{\text{Outside Work} \\ \text{Happens}}} \quad = \underbrace{KE_2 + PE_2}_{\text{Final Energy}} .$$

Note that this equation can be used in multiple locations in between, not just initial and final. We can break problems apart into stages, and this can be a very useful solution strategy in many problems.

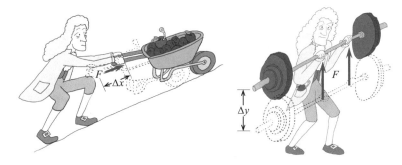

Figure 8.2: Newtdog works by applying forces in the direction of motion (©E. Diehl).

Energy methods are powerful tools to solve many kinds of engineering problems, often as an alternative to other solution techniques that become excessively complicated. We'll see that the kinetic energy change is closely related to N2L since it's derived from it. We'll start the derivation/explanation with the definition of work.

8.2   WORK (U)

What is the definition of "work"? You might answer with an equation, but perhaps a layman's answer like "effort to move stuff" or "getting stuff done" is more descriptive. As a technical definition we might phrase this as "work is force *through* a distance." There is emphasis on the word "through" because this is a necessary part of the concept. Work requires both force and movement, and only when the force and movement align is work done. In Figure 8.2, Newtdog is pushing the loaded wheel barrow up the hill. He's specifically pushing the handles in the direction of motion. The force he's exerting on the wheel barrow multiplied by the distance it moves in the direction of the force is the work he is applying. We can think of Work as external effort, but the aligning of force and movement is key. Work is its own type of energy which can be transformed into other types of energy.

A formal definition of that incremental work is the "scalar product" (a.k.a. "dot product") of the applied force and the incremental movement. The force and the movement are both vectors, but taking the scalar product removes the vector nature of both resulting in a scalar value (thus the name "scalar product"). We use the letter "U" rather than "W" to avoid confusion with the variable used for weight

$$dU = \overline{\mathbf{F}} \cdot d\overline{\mathbf{r}}.$$

Recall other consequential aspects of the scalar product include the "projection" of one vector onto another. Figure 8.3 shows a force applied to a particle that is moved from points 1 to 2 along a path. The work is the magnitude of the force times the magnitude of the displacement

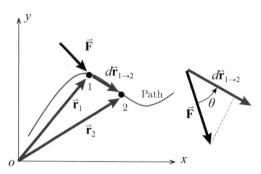

Figure 8.3: Work is the projection of the force on the change in position.

times the cosine of the angle between the vectors:

$$dU = |\mathbf{\vec{F}}|\,|d\mathbf{\bar{r}}| \cos \theta.$$

The important thing to note here is that energy doesn't have a direction after this dot product. This is especially important since although the Work-Energy method can simplify (often by reducing algebra and/or trigonometry), the results don't give us direction.

Taking the integral of the incremental work equation between two points gives us the work accomplished:

$$\boxed{U_{1\rightarrow 2} = \int_{1}^{2} \mathbf{\bar{F}} \cdot d\mathbf{\bar{r}}}.$$

If the force doesn't change with position (is constant) then: $U_{1\rightarrow 2} = F\Delta x$ (again, only if F and Δx align).

Work that is dependent on the path is said to be caused by a "non-conservative force." We'll see there are at least two common instances where we have "conservative forces," meaning that the path taken between two points doesn't matter (in other words it's path independent). We make a special case for this kind of work and call it "potential energy" as we'll discuss in an upcoming section.

Another important thing to note about the dot product is the scenario when the angle between vectors is $\theta = 180°$. This means, of course, the force is in the opposite direction of the movement. The result is a negative value of work. An example of this is friction. *Negative work means energy is leaving due to an external force (such as friction).*

8.3   KINETIC ENERGY (KE)

Newton's Second Law tells us that the net applied force produces motion when not zero. That motion has energy and applying a force can increase that motion energy. We'll hold off giving it a name until we arrive at its definition.

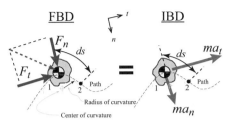

Figure 8.4: FBD/IBD in path coordinates moving from points 1 to 2.

In Figure 8.4, we revisit the FBD/IBD pair used in path coordinates and follow the path from positions 1 to 2 to track the energy exerted. The tangential component of the force travels along the path, so the work is:

$$\sum F_t = ma_t.$$

Following this path must be done in small increments, so we setup N2L in path coordinates with differentials. We know that $a_t = \frac{dv}{dt}$ but also $a_t = v\frac{dv}{ds}$.

$$F_t = mv\frac{dv}{ds}$$

$$F_t ds = mvdv$$

$$\int_1^2 F_t ds = \int_1^2 mvdv = \frac{1}{2}mv_2{}^2 - \frac{1}{2}mv_1{}^2.$$

The force travels along the path, so it is always in the direction of motion, therefore $U_{1\to2} = \int_1^2 F_t ds$. We call the motion energy caused by this work a change in "Kinetic Energy" (ΔKE):

$$\boxed{KE = \frac{1}{2}mv^2}.$$

The important take-away here is the link between N2L and Work-Energy.

8.4   POTENTIAL ENERGY (PE)

We can think of "Potential Energy" (PE) as a special type of Work where the path doesn't matter, only the location where it's measured. Other ways to phrase this are "path independent" or, as previously mentioned, work from "conservative forces." There are two common types of potential energy we'll use: "gravitational potential energy" (PE_g) and "spring potential energy" (PE_{sp}). We can also think of potential energy as "stored energy." There are other kinds of potential energy (chemical potential energy for instance) but these are the ones we use in dynamics.

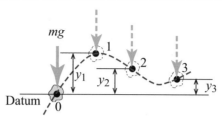

Figure 8.5: Gravitational potential energy depends on vertical position with respect to a datum.

8.4.1   GRAVITATIONAL POTENTIAL ENERGY (PE_g)

The type of work energy produced by gravitational forces is called "gravitational potential energy" (PE_g). As shown in Figure 8.5, the force of the work energy, weight (mg), always points downward.

All changes in position aligned with the weight force are vertical so $U_{1\rightarrow 2} = -mg\Delta y_{1\rightarrow 2} = -mgy_2 + mgy_1$. This is negative work because the force is in the opposite direction to the change in position. We define gravitational potential energy as:

$$\boxed{PE_g = mgy = Wy}\,.$$

Note we can use mg and W interchangeably. Again, note that we chose the letter "U" for Work rather than "W" to avoid any confusion there might be with weight.

To keep track of gravitational potential energy we need to establish a datum (reference) to measure the vertical position. We'll often find it convenient to set one of the potential energies to zero, to avoid negative potential energy. To keep track of this it's best to keep in mind whether the stored energy has increased or decreased between states. For instance, when PE_{g1} is larger than PE_{g2} we might choose PE_{g2} to be equal to zero, thereby making it the datum. Or, since we have another energy state at point 3 later, we might choose to make PE_{g3} zero as the datum.

8.4.2   SPRING POTENTIAL ENERGY (PE_{sp})

Springs also store energy when they're deformed either by compression or extension. We will limit our discussion to linear springs where force and deflection are proportional by a constant spring rate, also called "stiffness" (k). The deflection is the difference between the undeformed length and the current length:

$$\delta_1 = l_0 - l_1 \quad \text{and} \quad \delta_2 = l_0 - l_2.$$

On the plot of force vs. deflection in Figure 8.6, stiffness is represented by the slope of the straight line (and therefore "linear").

The force required to stretch a spring to any deflection, δ, is:

$$F = k\delta.$$

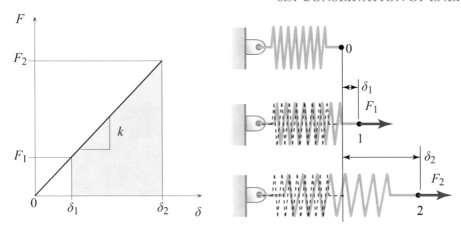

Figure 8.6: Spring force vs. deflection at two stretched positions.

The work required to stretch the spring from points 1 to 2 is:

$$U_{1\to 2} = \int_1^2 F d\delta = \int_1^2 (k\delta)d\delta = \frac{1}{2}k\delta_2{}^2 - \frac{1}{2}k\delta_1{}^2.$$

This work is stored energy and is also the area under the curve of the plot in Figure 8.6. We define "spring potential energy" at any state as:

$$\boxed{PE_{sp} = \frac{1}{2}k\delta^2}.$$

We're careful to avoid this common mistake: $\Delta PE_{sp} = \frac{1}{2}k\left(\delta_2{}^2 - \delta_1{}^2\right) \neq \frac{1}{2}k(\delta_2 - \delta_1)^2$.

To represent deflection, the variable lowercase Greek delta (δ) is purposefully used here instead of the letter "x" used by many texts. This is to emphasize that it isn't dependent on direction but is dependent on change in length as depicted in Figure 8.7.

8.5   CONSERVATION OF ENERGY

When there is no external work, only kinetic and potential energy changes, we apply the Conservation of Energy principal, written as:

$$\boxed{KE_1 + PE_1 = KE_2 + PE_2}.$$

This situation often occurs when a system is frictionless and no external forces are applied in line with the motion. For example, normal forces can be acting on an object but don't constitute external work because there is no motion in the normal direction. We can see from the above

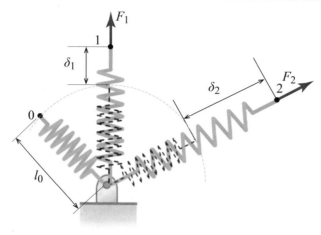

Figure 8.7: Spring deflection is independent of direction.

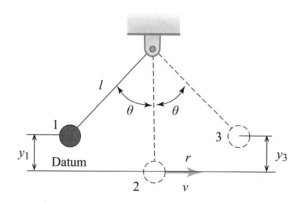

Figure 8.8: Pendulum demonstrating conservation of energy.

equation that conservation of energy is a trade off between kinetic energy and potential energy just like the apple in Figure 8.1. Consequently, kinetic energy is at a maximum when potential energy is at a minimum and vice versa.

A pendulum is an excellent example of this concept, as shown in Figure 8.8.

In position 1, the ball is released from rest. Before it moves it has no kinetic energy, only potential. When it reaches the bottom, position 2, it is at its maximum velocity and therefore maximum kinetic energy but has reached the datum so has zero potential energy. When it rises to position 3, it again momentarily stops moving (no kinetic energy) but has regained the potential energy it had in position 1. This depends on there being no external losses such as air resistance on the ball or friction at the pivot point. In equations we'd write this as:

$$PE_1 = KE_2 = PE_3$$

$$mgy_1 = \frac{1}{2}mv_2{}^2 = mgy_3.$$

We'll note that the masses will cancel which is an interesting aspect of pendula (their motion is independent of mass). We also conclude that the ball will return to the original height, $y_1 = y_3$, as long as there are no energy losses.

8.6   SOLVING WORK-ENERGY METHOD PROBLEMS

Applying the Work-Energy Method typically involves some typical steps.

1. If there is friction in the problem, an FBD/IBD should be drawn and the normal forces calculated in order to find the friction force.

2. If there is dependent motion in the problem, the relations (position changes and velocity) should be calculated. This step and step 1 can be performed later as an aside, but if we recognize they're necessary it's good to get them out of the way.

3. Recreate a sketch of the problem and draw dashed line versions of the object(s) in the new positions/states to be analyzed. If the overall sketch is too cluttered to make this effective, draw several sketches.

4. Label each state as 1, 2, 3… Often there will only be two states, but some problems require more stages of movement, especially when one of the forms of energy no longer changes. An example of this might be two blocks attached by a rope and one of the blocks reaches the ground while the other keeps moving (see Example 8.4).

5. Establish a datum for zero potential energy at the lowest state and calculate the vertical distances from it to each stage. If there is more than one object a datum can be established for each one to avoid using negative potential energy. We may wish to reset the datum if more than 2 states are needed in the problem.

6. Write out the Work-Energy equation for the first stage: $KE_1 + PE_1 + U_{1\to2} = KE_2 + PE_2$.

7. List the forms of energy as KE_1, PE_1, $U_{1\to2}$, KE_2, and PE_2, and write out the equations for each as they apply, setting several to zero where appropriate. Note that when $U_{1\to2} = 0$ the problem is Conservation of Energy.

8. Enter in the knowns and identify the unknown parameter(s)

9. If necessary, apply the dependent motion from step 2 to link unknowns and solve.

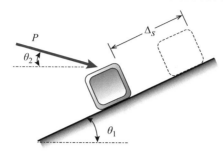

Figure 8.9: Example 8.1, repeat of Figure 6.6.

10. If necessary repeat 6–9 for each additional state.

Note that if acceleration or time is asked (or provided as a known in the problem), we will need to apply kinematics since neither parameter is a variable in Work-Energy. It's also sometimes necessary to use N2L to assist with forces. We can only use constant acceleration kinematics when the forces are constant. For example, when springs are in the system force varies throughout the change in position so the acceleration is not constant.

We'll demonstrate the application of Work-Energy on examples we've already solved using N2L. We'll need to use some kinematics to get results we can compare because, as we'll discuss further in Impulse-Momentum, Work-Energy is well suited to problems that ask for velocities and displacements, while N2L is well suited for problems involving accelerations. In the next class we will demonstrate even more complicated examples as well as introduce the concept of efficiency and power.

We are still dealing with Newtonian Dynamics, but it's good to be aware that other types of Dynamics also use energy methods (especially Hamiltonian and Lagrangian) and can circumvent tedious or difficult algebra and trigonometry for problems more complicated than covered in this text.

Example 8.1 (This is a repeat of Example 6.1.)
A constant $P = 50$ lb force is applied to a box weighing 25 lb, starting from rest, and positioned on a $\theta_1 = 25°$ inclined surface with $\mu_s = 0.25$ and $\mu_k = 0.2$ static and kinetic coefficients of friction, respectively (Figure 8.9). The force is applied to box $\theta_2 = 15°$ from horizontal. Determine the distance up the slope box travels (Δs) when it reaches a speed of $v = 3$ ft/s.

We'll walk through this problem labeling steps for demonstration of a typical Work-Energy Method procedure.

Step 1: FBD/IBD to determine the normal force and from that the friction force (Figure 8.10):

$$\nwarrow \sum F_{y'} = ma_{y'} = 0$$
$$- P \sin (\theta_1 + \theta_2) - W \cos \theta_1 + F_N = 0$$

FBD   IBD

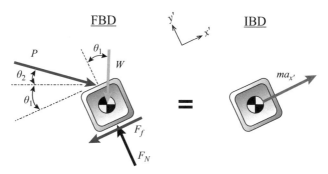

Figure 8.10: FBD/IBD of Example 8.1, repeat of Figure 6.7.

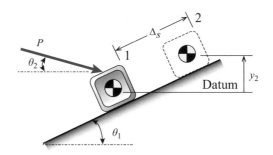

Figure 8.11: Example 8.1 work-energy stages.

$$F_N = (50)\sin(40°) + (25)\cos(25°) = 54.80 \text{ lb}$$

$$F_f = \mu_k F_N = (0.20)(54.80) = 10.96 \text{ lb}.$$

Step 2: No dependent motion, so skip.

Steps 3 and 4: Redraw problem with labels of positions and datum (Figure 8.11).

Step 5: Calculate stage vertical distances from datum $\quad y_2 = \Delta s \sin\theta_1$.

Step 6: Write out Work-Energy Equation

$$KE_1 + PE_1 + U_{1\rightarrow 2} = KE_2 + PE_2.$$

Step 7: Write out each energy as a list

$\quad KE_1 = 0$ Starts at rest

$\quad PE_1 = 0$ At datum

$U_{1\to2} = P\cos(\theta_1 + \theta_2)\,\Delta s - F_f\,\Delta s$. Only the portion of the external force that's in the direction of motion causes work. The work due to friction is negative because the friction force is in the opposite direction of motion

$$KE_2 = \tfrac{1}{2}mv_2{}^2$$

$$PE_2 = mgy_2 = Wy_2 = W\Delta s \sin\theta_2.$$

Step 8: Enter knowns

$$(0) + (0) + P\cos(\theta_1 + \theta_2)\,\Delta s - F_f\,\Delta s = \frac{1}{2}mv_2{}^2 + W\Delta s \sin\theta_1$$

$$(0) + (0) + \underbrace{(50)\cos\left((25°) + (15°)\right)}_{38.30}\Delta s - (10.96)\,\Delta s = \underbrace{\frac{1}{2}\left(\frac{25}{32,2}\right)(3)^2}_{3.494} + \Delta s\,\underbrace{(25)\sin(25°)}_{10.57}.$$

We identify Δs is the only unknown so it can be solved.

$$\Delta s = 0.2083 \text{ ft} = 2.500 \text{ in}$$

$$\boxed{\Delta s = 2.50 \text{ in}}.$$

Steps 9 and 10 aren't necessary in this problem.

Checking back to Example 6.1 we see the solutions match, and we can conclude that either approach would work for this problem. We will also attempt to use Impulse-Momentum on this example.

Example 8.2 (This is a repeat of Example 6.3)

The three setups shown begin at rest. In setup (a) the force ($P = 50$ lb) is applied to the cable attached to block A ($W_A = 75$ lb). Setup (b) has the same block A and is connected to block B ($W_B = 50$ lb). Setup (c) has larger blocks with the same difference between them ($W_A = 175$ lb and $W_B = 150$ lb) (Figure 8.12). The pulleys are assumed to be massless and frictionless. Determine for each setup the acceleration of block A.

This example demonstrates using multiple particles (objects) as well as allowing us to compare using N2L to Work-Energy.

This problem asks for acceleration but Work-Energy deals with velocity and displacement (Figure 8.13). In order to compare the results to Example 6.3, we'll need to make an assumption. Let's pick a displacement and determine the velocity. From this we can use kinematics to determine the acceleration. How do we know to do this? We look at the situation and compare it to the concept, see an obstacle and search for a way around it. Let's proceed as if we didn't already know our solution strategy and see where we get stuck.

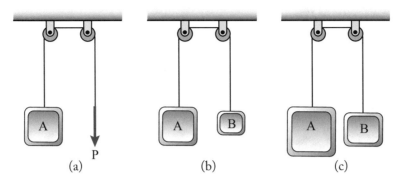

Figure 8.12: Example 8.2, repeat of Figure 6.11.

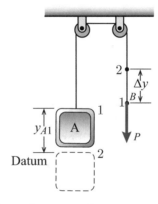

Figure 8.13: Example 8.2 Part (a) work-energy diagram.

Step 1: There is no friction so we don't need this step.

Step 2: The dependent motion can be established from observation: $\Delta y_A = |\Delta y_B|$ and $v_A = |v_B|$.

Steps 3 and 4: Redraw and label (Figure 8.13).

Step 5: We don't know the change in position, but will assume one as mentioned above. Let's just keep these as variables and apply $y_{A2} = |\Delta y|$.

Step 6: Write out Work-Energy equation

$$KE_1 + PE_1 + U_{1 \to 2} = KE_2 + PE_2.$$

Step 7: Write out each energy as a list

$KE_1 = 0$ Starts at rest

$PE_1 = W_A y_{A1}$ A begins above the datum, B has no PE

$U_{1\rightarrow 2} = -P\Delta y$ The external force is in the opposite direction of the motion so this is negative external work

$KE_2 = \frac{1}{2} m_A v_{A2}{}^2$ Only A has KE

$PE_2 = 0$ A finishes at the datum

Step 8: Enter knowns

$$(0) + W_A y_{A1} - P\Delta y = \frac{1}{2} m_A v_{A2}{}^2 + (0).$$

We identify y_{A2}, Δy, and v_{A2} as the unknowns. Here is where we would likely get stuck. To proceed, we consider what we don't know (y_{A2}, Δy, and v_{A2}) and what we're trying to calculate (acceleration). This is where we should recognize that constant acceleration kinematics could help us: $v^2 = (v_0)^2 + 2a\Delta y$. We know we can use it because the force is constant and therefore the acceleration is constant. Let's assume a displacement of $\Delta y = 1$ ft, find the velocity from Work-Energy and then find the acceleration from that. We can test out our strategy by trying a different Δy to see if we get the same acceleration:

$$(0) + (75)(1) - (50)(1) = \frac{1}{2}\left(\frac{75}{32.2}\right) v_{A2}{}^2 + (0)$$

$$v_{A2} = 4.633 \text{ ft/s}.$$

Using constant acceleration kinematics: $v^2 = (v_0)^2 + 2a\Delta y$

$$(4.633)^2 = (0)^2 + 2a(1)$$

$$a_A = 10.73 \text{ ft/s}^2$$

(a) $\boxed{\vec{a}_A = 10.73 \text{ ft/s}^2 \ \downarrow}$.

We compare this back to Example 6.3 and see we get the same result.

We know the direction based on the setup of the problem, not the signs in the answer. This is because the velocity in Kinetic Energy is squared, so the direction of velocity can't be established from sign conventions as is can in N2L. This might be considered a weakness of the Work-Energy Method which also holds true for other types of Dynamics that rely on energy principles.

We'll finish the remainder of the examples without labeling steps (Figure 8.14).

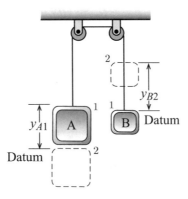

Figure 8.14: Example 8.2 part (b) work-energy diagram.

Part (b)

$$KE_1 + PE_1 + U_{1\to 2} = KE_2 + PE_2,$$

where

$KE_1 = 0$ Starts from rest

$PE_1 = W_A y_{A1}$ A starts above its datum, B starts at its datum

$U_{1\to 2} = 0$ No external work, (conservation of energy)

$KE_2 = \frac{1}{2}m_A v_{A2}^2 + \frac{1}{2}m_B v_{B2}^2$

$PE_2 = W_B y_{B2}$ A finishes at the datum

There is no external work in scenario (b) as there was in (a), and there are two datums, one for each block:

$$(0) + W_A y_{A1} + (0) = \frac{1}{2}m_A v_{A2}^2 + \frac{1}{2}m_B v_{B2}^2 + W_B y_{B2}$$

$$(0) + (75)\, y_{A1} + (0) = \frac{1}{2}\left(\frac{75}{32.2}\right) v_{A2}^2 + \frac{1}{2}\left(\frac{50}{32.2}\right) v_{B2}^2 + (50)\, y_{B2}.$$

Set $v_{B2} = v_{A2}$ and $y_{A1} = y_{B2} = 1$ ft

$$(0) + (75)\,(1) + (0) = \frac{1}{2}\left(\frac{75}{32.2}\right) v_{A2}^2 + \frac{1}{2}\left(\frac{50}{32.2}\right) v_{A2}^2 + (50)\,(1)$$

$$v_{A2} = 3.589 \text{ ft/s}.$$

Apply kinematics: $v_{A2}^2 = (v_{A1})^2 + 2a_A \Delta y_A$

$$(3.589)^2 = (0)^2 + 2a_A\,(1)$$

$$a_A = 6.440 \text{ ft/s}^2$$

(b) $\boxed{\vec{a}_A = 6.44 \text{ ft/s}^2 \downarrow}$.

Part (c) We can reuse most of part (b) because the only differences are the weights:

$$(0) + W_A y_{A1} + (0) = \frac{1}{2} m_A v_{A2}^2 + \frac{1}{2} m_B v_{B2}^2 + W_B y_{B2}$$

$$(0) + (175) y_{A1} + (0) = \frac{1}{2} \left(\frac{175}{32.2} \right) v_{A2}^2 + \frac{1}{2} \left(\frac{150}{32.2} \right) v_{B2}^2 + (150) y_{B2}.$$

Set $v_{B2} = v_{A2}$ and $y_{A1} = y_{B2} = 1$ ft

$$(0) + (175)(1) + (0) = \frac{1}{2} \left(\frac{175}{32.2} \right) v_{A2}^2 + \frac{1}{2} \left(\frac{150}{32.2} \right) v_{A2}^2 + (150)(1)$$

$$v_{A2} = 2.226 \text{ ft/s}.$$

Apply kinematics: $v_{A2}^2 = (v_{A1})^2 + 2a_A \Delta y_A$

$$(2.226)^2 = (0)^2 + 2a_A (1)$$

$$a_A = 2.477 \text{ ft/s}^2.$$

(c) $\boxed{\vec{a}_A = 2.48 \text{ ft/s}^2 \downarrow}$.

We see that parts (b) and (c) have the same results as found in Example 6.3. Remember to consider the implications of these results: mass causes the motion to be more sluggish. We can also appreciate the exchange of potential energy and kinetic energy within this system.

Example 8.3   (This is a partial repeat of Example 7.2)
Newtdog has tied Wormy's large apple ($W = 0.5$ lb) to a 2 ft string to make a pendulum. He starts it in motion, and the maximum speed (when the apple is at the bottom of the arc) is 1 ft/s (Figures 8.15 and 8.16). What is the maximum angle it will rise (measured from the vertical axis)? (Note: we are not asking for the tension in the string.)
   Vertical position at top of swing: $y_2 = l - l \cos \theta = l (1 - \cos \theta)$

$$KE_1 + PE_1 + U_{1 \to 2} = KE_2 + PE_2,$$

where

$$KE_1 = \tfrac{1}{2} m v_1^2 \quad \text{At maximum speed}$$

$$PE_1 = 0 \quad \text{At datum}$$

Figure 8.15: Example 8.3, repeat of Figure 7.6 (©E. Diehl).

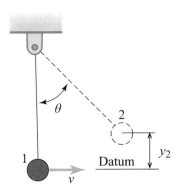

Figure 8.16: Example 8.3 Work-Energy diagram.

$U_{1 \to 2} = 0$ No external forces, therefore conservation of energy

$KE_2 = 0$ Ball pauses at top of swing

$PE_2 = mgy_2 = mgl\,(1 - \cos\theta)$

$$\frac{1}{2}mv_1{}^2 + (0) + (0) = (0) + mgl\,(1 - \cos\theta).$$

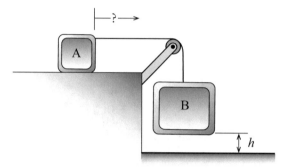

Figure 8.17: Example 8.4.

Note that the mass cancels out

$$\frac{1}{2}v_1{}^2 + (0) + (0) = (0) + gl\,(1 - \cos\theta)$$

$$\frac{1}{2}(1)^2 + (0) + (0) = (0) + (32.2)\,(2)\,(1 - \cos\theta)$$

$$\theta = 71.44°$$

$$\boxed{\theta = 71.4°}\,.$$

This agrees with the results from Example 7.2 (b). This is a good example of when solving a problem using Work-Energy is much easier than using N2L. On the other hand, finding the maximum string tension would be much more difficult using Work-Energy since magnitude and direction of that force is not constant throughout the motion nor does it align with the motion.

Example 8.4
Two blocks are connected by a rope over a massless and frictionless pulley (Figure 8.17). The bottom of block B ($m = 60$ kg) is $h = 0.5$ m above the ground. Block A ($m = 35$ kg) is on a horizontal surface with a coefficient of friction of $\mu_k = 0.30$. The blocks are released from rest and begin to move. Determine the distance block A will travel before coming to rest. This problem demonstrates a scenario where more than two states are needed.

Blocks A and B will move together until block B hits the ground and the rope becomes slack. The initial position is state 1, when block B hits the ground is state 2, and when Block A stops is state 3.

We should find the friction force on block A using an FBD/IBD pair in Figure 8.18. Note that we don't technically have to use the IBD since we'll only use the vertical direction which has no acceleration, but it's good practice to not take short cuts.

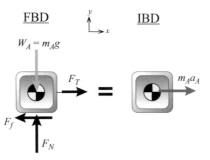

Figure 8.18: Example 8.4 FBD/IBD pair.

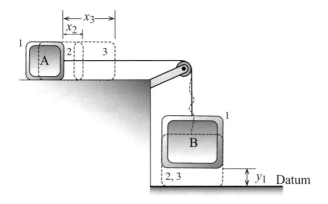

Figure 8.19: Example 8.4 Work-Energy diagram.

$$\uparrow \sum F_y = ma_y = 0$$

$$- m_A g + F_N = 0$$

$$F_N = (35)\,(9.81) = 343.4 \text{ N}$$

$$F_f = \mu_k F_N = (0.30)\,(343.4) = 103.0 \text{ N.}$$

States $1 \Rightarrow 2$ (Figure 8.19)

$$KE_1 + PE_1 + U_{1 \to 2} = KE_2 + PE_2$$

where

$$KE_1 = 0 \quad \text{Starts at rest}$$

$$PE_1 = m_B g y_1 \quad \text{block } B \text{ is above datum}$$

$U_{1\to2} = -F_f x_2$ Negative because friction force is opposite from movement

$KE_2 = \frac{1}{2}m_A v_{A2}^2 + \frac{1}{2}m_B v_{B2}^2$ Blocks kinetic energy the instant before block B hits the ground

$PE_2 = 0$ block B is at datum

$$(0) + m_B g y_1 - F_f x_2 = \frac{1}{2}m_A v_{A2}^2 + \frac{1}{2}m_B v_{B2}^2 + (0).$$

The relations we can use are: $x_2 = y_1 = 0.5$ m and $v_{A2} = v_{B2}$

$$(0) + \underbrace{(60)\,(9.81)\,(0.5)}_{294.3} - \underbrace{(103.0)\,(0.5)}_{51.51} = \frac{1}{2}\,(35)\,v_{A2}^2 + \frac{1}{2}\,(60)\,v_{A2}^2 + (0)$$

$$v_{A2} = 2.261 \text{ m/s.}$$

Note that the friction energy loss *could* be enough to stop block B from reaching the bottom. This would present itself as having to take the square root of a negative number (look at the $m_B g y_1$ vs. $F_f \Delta x_2$ intermediate values).

States $2 \Rightarrow 3$

$$KE_2 + PE_2 + U_{2\to3} = KE_3 + PE_3$$

where

$KE_2 = \frac{1}{2}m_A v_{A2}^2$ Only block A's kinetic energy at stage 2

$PE_2 = 0$ No potential energy change for block A

$U_{2\to3} = -F_f\,(x_3 - x_2)$ Friction force times distance traveled between stages 2 and 3

$KE_3 = 0$ Block A comes to rest

$PE_3 = 0$ No potential energy change for block A

$$\frac{1}{2}m_A v_{A2}^2 + (0) - F_f\,(x_3 - x_2) = (0) + (0)$$

$\Delta x_2 = 0.5$ m

$$\frac{1}{2}\,(35)\,(2.261)^2 + (0) - (103.0)\,(x_3 - (0.5)) = (0) + (0)$$

$$x_3 = 1.368 \text{ m}$$

$$\boxed{\Delta x_A = 1.37 \text{ m}}.$$

Could this have been found directly from states 1 to 3?

$$KE_1 + PE_1 + U_{1\to3} = KE_3 + PE_3,$$

where

$KE_1 = 0$ Starts at rest

$PE_1 = m_B g y_1$ Block B is above datum

$U_{1\to3} = -F_f x_3$ Friction force times total distance traveled

$KE_3 = 0$ Blocks kinetic energy the instant before block B hits the ground

$PE_3 = 0$ Block B is at datum

$$(0) + \underbrace{(60)\,(9.81)\,(0.5)}_{294.3} - (103.0)\,x_3 = (0) + (0)$$

$x_3 = 2.857\ m$.... Not the same...why? The kinetic energy in B is lost when it hits the ground and the rope goes slack. Recognizing that more than two states are necessary is the key to solving this problem.

In order to do this problem using N2L you'd need to take FBD/IBD pairs of each block to find the accelerations linked by the rope tension and then use kinematics to find the velocity of A when B strikes the ground. Then you'd recalculate the deceleration of block A alone and use kinematics to find how far it would travel to reach zero speed.

Example 8.5
A block ($m = 1$ kg) connected to two springs (both $k = 100$ N/m) rides on a frictionless rod in the vertical plane. In the position shown (Figure 8.20) the springs are in their undeformed lengths. The dimensions are $a = 0.6$ m, $b = 0.9$ m, and $c = 0.8$ m. The block is raised into position ① and released. Determine the block's speed when it strikes D in position ③.

From Figure 8.21 we see that spring AB is stretched to length l_1 at state 1 and l_3 at state 3. We also see that spring BC is compressed at state 1. We can find the deflections of each by the difference from the undeformed length:

$$\delta_{AB1} = l_1 - c = \sqrt{a^2 + c^2} - c = \sqrt{(0.6)^2 + (0.8)^2} - (0.8) = 0.2000\ \text{m}$$

$$\delta_{BC1} = a = 0.6\ \text{m}$$

$$\delta_{AB1} = l_3 - c = \sqrt{b^2 + c^2} - c = \sqrt{(0.9)^2 + (0.8)^2} - (0.8) = 0.4042\ \text{m}.$$

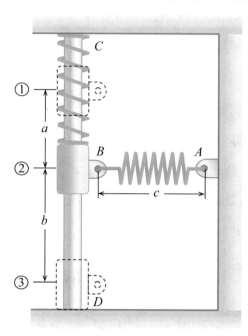

Figure 8.20: **Example** 8.5.

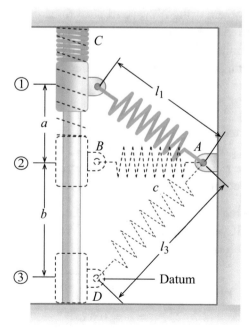

Figure 8.21: **Example** 8.5 Work-Energy diagram.

States $1 \Rightarrow 2$

$$KE_1 + PE_1 + U_{1\to2} = KE_2 + PE_2$$

where

$KE_1 = 0$ Starts at rest

$PE_{g1} = mgy_1 = mg\,(a+b)$ Block is above datum

$PE_{sp1} = \frac{1}{2}k\delta_{AB1}^2 + \frac{1}{2}k\delta_{BC1}^2$ Spring AB is stretched and spring BC is compressed

$U_{1\to2} = 0$ No external forces, conservation of energy

$KE_2 = \frac{1}{2}mv_2^2$ Kinetic energy at state 2

$PE_{g2} = mgy_2 = mgb$ Block is still above datum

$PE_{sp2} = 0$ Springs are in undeformed

$$(0) + mg\,(a+b) + \frac{1}{2}k\delta_{AB1}^2 + \frac{1}{2}k\delta_{BC1}^2 + (0) = \frac{1}{2}mv_2^2 + mgb + (0)$$

$$(0) + \underbrace{(1)\,(9.81)\,((0.6)+(0.9))}_{14.72}$$

$$+ \underbrace{\frac{1}{2}\,(100)\,(0.2000)^2}_{2.000} + \underbrace{\frac{1}{2}\,(100)\,(0.6)^2}_{18.00} + (0)$$

$$= \frac{1}{2}\,(1)\,v_2^2 + \underbrace{(1)\,(9.81)\,(0.9)}_{8.829} + (0)$$

$$v_2 = 7.196 \text{ m/s.}$$

States $2 \Rightarrow 3$

$$KE_2 + PE_2 + U_{2\to3} = KE_3 + PE_3$$

where

$KE_2 = \frac{1}{2}mv_2^2$ Kinetic energy at state 2

$PE_{g2} = mgy_2 = mgb$ Block is still above datum

$PE_{sp2} = 0$ Springs are undeformed

$U_{2\to3} = 0$ No external forces, conservation of energy

$KE_3 = \frac{1}{2}mv_3^2$ Kinetic energy just before block hits the bottom

$PE_{g3} = 0$ Block is at datum

$PE_{sp1} = \frac{1}{2}k\delta_{AB3}{}^2$ Spring AB is stretched

$$\frac{1}{2}mv_2{}^2 + mgb + (0) + (0) = \frac{1}{2}mv_3{}^2 + (0) + \frac{1}{2}k\delta_{AB3}{}^2$$

$$\underbrace{\frac{1}{2}(1)(7.196)^2}_{25.89} + \underbrace{(1)(9.81)(0.9)}_{8.826} + (0) = \frac{1}{2}(1)v_3{}^2 + (0) + \underbrace{\frac{1}{2}(100)(0.4042)^2}_{8.169}$$

$$v_3 = 7.287 \text{ m/s.}$$

Could this have been found directly from states 1 to 3?

$$KE_1 + PE_1 + U_{1\rightarrow3} = KE_3 + PE_3$$

$$(0) + (14.72) + (2.000) + (18.00) + (0) = \frac{1}{2}(1)v_3{}^2 + (0) + (8.169)$$

$$v_3 = 7.287 \text{ m/s} \quad \text{same answer!}$$

We get the same answer going from state 1 to 3 as we did from going from 1 to 2 and then 2 to 3. Why? We note that both changes in state are conservation of energy, and unlike the previous example no energy was removed as was the case when block B's kinetic energy in Example 8.4 was no longer contributing to the motion. So we are able to skip state 2 entirely.

You may have noted that the values for different energies were calculated and included beneath. This is a useful step many students leave off, but it can help problem solving in several ways. First, this assists trouble-shooting and checking for numerical errors when you arrive at an answer that doesn't make physical sense. Second, finding these in-between values helps contextualize the contribution of each energy type. Next, we sometimes will need to reuse values, for instance when checking if we could go from states 1–3. Last, it helps the grader of homework and exams locate simple math mistakes. This can help you get more partial credit since most graders don't remove as many points for numerical mistakes as conceptual mistakes.

This example is included because none of the other examples used springs.

This problem would become significantly more difficult if friction were included as the normal force of the block on the rod would change with the spring orientation. Work-Energy might still be preferable to N2L in this revised scenario. Springs that change orientation can make using N2L with kinematics difficult as not only do the forces change with position but also acceleration is not constant.

In the next class we will do more Work-Energy examples with a variety of added complexities.

Book 1 - Class 9

https://www.youtube.com/watch?v=J3fciZ3eOAU

CLASS 9

Work-Energy Method and the Conservation of Energy (Part 2)

B.L.U.F. (Bottom Line Up Front)

- Work-Energy: $KE_1 + PE_1 + U_{1\to 2} = KE_2 + PE_2$.

- Conservation of Energy: $KE_1 + PE_1 = KE_2 + PE_2$.

- Efficiency: $\eta = \frac{energy\ output}{energy\ input} = \frac{WYW}{WYPF}$.

- Power: $\mathbb{P} = \frac{dU}{dt} = F \cdot \frac{dr}{dt} = F \cdot v$.

9.1   WORK-ENERGY PROBLEMS

In the previous class we introduced the Work-Energy method with this basic formula:

$$\underbrace{KE_1 + PE_1}_{Initial\ Energy} + \underbrace{U_{1\to 2}}_{\substack{External \\ Energy\ Change}} = \underbrace{KE_2 + PE_2}_{Final\ Energy} \ .$$

Without the external energy change, this reduces to the Conservation of Energy principle:

$$KE_1 + PE_1 = KE_2 + PE_2.$$

We also demonstrated that in many instances Work-Energy combined with kinematics could be used instead of N2L to get the same solution. Since the topic is important, a second class/chapter is included here to demonstrate a few more complex examples.

Types of Work-Energy Problems:

- Conservation of Energy: balance between potential and kinetic energy (no external energy)

- External Work Applied

- Frictional Energy Loss

- Multiple Particles

- Power and Efficiency

9.2   EFFICIENCY

The first law of thermodynamics can be phrased in a variety of ways, including "energy cannot be created or destroyed" which is also the Conservation of Energy principle we've been using. A fun way to say this is "you can't get something for nothing." Along the same lines, we can phrase the second law of thermodynamics as "you can't even break even." It's this second law that leads us to the topic of "efficiency": the ratio of output to input. The second law of thermodynamics tells us that ratio will always be less than 1. We can define this mathematically using the lower case Greek letter "eta" as:

$$\eta = \frac{energy\ output}{energy\ input}.$$

In the spirit of the fun way to phrase things, we can use the following to remember how to calculate efficiency:

$$\eta = \frac{what\ you\ want}{what\ you\ paid\ for} = \frac{WYW}{WYPF}.$$

If we perform a simple energy balance we see: $energy\ input - energy\ loss = energy\ output$. So we can rewrite the efficiency in terms of losses:

$$\eta = 1 - \frac{energy\ loss}{energy\ input}.$$

Example 9.1

A spring ($k = 25$ kN/m) is used to stop a $m = 75$ kg package that is moving down a $\theta = 60°$ incline with initial speed $v_0 = 3$ m/s when it is $d = 10$ m from the spring (Figure 9.1). The coefficient of kinetic friction between the package and the incline is $\mu_k = 0.20$. What is the maximum distance up the incline (from the spring) the package will travel after it bounces off the spring? What is the efficiency of this process?

We need N2L for the normal force in the y' direction so we can find the friction force from an FBD/IBD pair (Figure 9.2):

$$\nwarrow \sum F_{y'} = ma_{y'} = 0$$

$$- mg \cos \theta + F_N = 0$$

$$F_N = (75)(9.81) \cos(60°) = 367.9\ \text{N}$$

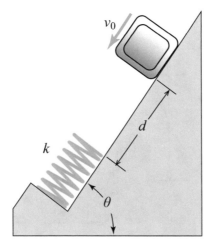

Figure 9.1: Example 9.1.

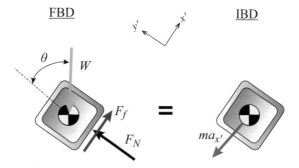

Figure 9.2: FBD/IBD pair of Example 9.1.

$$F_f = \mu_k F_N = (0.20)(367.9) = 73.58 \text{ N.}$$

This problem will benefit from solving it in three states. We choose state two as when the package comes to a stop at the spring's maximum deflection. Our goal in this is to find δ. We'll use this as the datum, using the bottom corner of the package as the reference (Figure 9.3).

States $1 \Rightarrow 2$

$$KE_1 + PE_1 + U_{1\to2} = KE_2 + PE_2.$$

The distance of state 1 from the datum is:

$$y_1 = (d + \delta)\sin\theta$$

$$KE_1 = \tfrac{1}{2}mv_0^2$$

$$PE_1 = mgy_1 = mg(d + \delta)\sin\theta$$

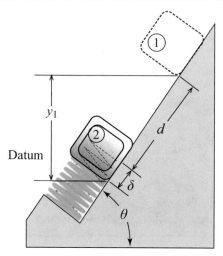

Figure 9.3: Example 9.1 Work-Energy diagram for states 1 to 2.

$$U_{1\to2} = -F_f \Delta x' = -F_f (d + \delta)$$

$$KE_2 = 0$$

$$PE_2 = \tfrac{1}{2}k\delta^2$$

$$\frac{1}{2}mv_0^2 + mg (d + \delta) \sin\theta - F_f (d + \delta) = (0) + \frac{1}{2}k\delta^2$$

$$\underbrace{\frac{1}{2} (75) (3)^2}_{337.5 \text{ Nm}} + \underbrace{(75) (9.81) ((10) + \delta) \sin (60°)}_{(6,372)+(637.2)\delta} - \underbrace{(73.58) (10 + \delta)}_{-(735.8)-(73.58)\delta} = (0) + \underbrace{\frac{1}{2} (25,000) \delta^2}_{(12,500)\delta^2}$$

$$(12,500)\,\delta^2 - (563.6)\,\delta - (5,974) = 0.$$

Find the roots with either the quadratic equation or a root solver:

$$\delta = \frac{-b \pm \sqrt{b^2 - 4ac}}{2a} = \frac{-(-563.6) \pm \sqrt{(-563.6)^2 - 4\,(12,500)\,(-5,974)}}{2\,(12,500)}$$

$$= 0.7142, \quad -0.6691 \text{ m}.$$

The negative answer is meaningless. So the deflection is $\delta = 0.7142$ m.

States $2 \Rightarrow 3$ (Figure 9.4)

$$KE_2 + PE_2 + U_{2\to3} = KE_3 + PE_3$$

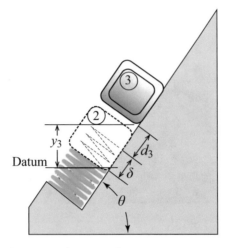

Figure 9.4: Example 9.1 work-energy diagram for states 2 to 3.

where

$$KE_2 = 0$$

$$PE_2 = \frac{1}{2}k\delta^2$$

$$U_{2\to3} = -F_f(d_3 + \delta)$$

$$KE_3 = 0$$

$$PE_3 = mgy_3 = mg(d_3 + \delta)\sin\theta$$

$$(0) + \frac{1}{2}k\delta^2 - F_f(d_3 + \delta) = (0) + mg(d_3 + \delta)\sin\theta$$

$$(0) + \underbrace{\frac{1}{2}(25{,}000)(0.7142)^2}_{(6{,}376)} - \underbrace{(73.58)(d_3 + (0.7142))}_{-(73.58)d_3-(52.55)}$$

$$= (0) + \underbrace{(75)(9.81)(d_3 + (0.7142))\sin(60°)}_{(637.2)d_3+(455.1)}$$

$$d_3 = 8.256 \text{ m}$$

$$\boxed{d_3 = 8.26 \text{ m}}.$$

This problem needed to be done in stages because the deflection of the spring is part of the distance within the frictional energy loss.

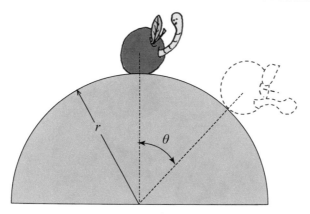

Figure 9.5: Example 9.2 (© E. Diehl).

The initial energy is the energy input:

$$KE_1 + PE_1 = \frac{1}{2}mv_0^2 + mg(d+\delta)\sin\theta$$

$$= \underbrace{\frac{1}{2}(75)(3)^2}_{337.5} + \underbrace{(75)(9.81)((10)+(0.7142))\sin(60°)}_{6,827} = 7{,}164 \text{ Nm}.$$

Energy losses:

$$Losses = F_f(d+\delta) + F_f(d_3+\delta)$$

$$= (73.58)(10+(0.7142)) + (73.58)((8.256)+(0.7142)) = 1{,}448 \text{ Nm}$$

$$\eta = 1 - \frac{energy\ loss}{energy\ input} = 1 - \frac{(1{,}448)}{(7{,}164)} = 79.79\%$$

$$\boxed{\eta = 79.8\%}.$$

Example 9.2

Wormy's apple ($m = 0.25$ kg) is on top of a smooth cylinder ($r = 0.5$ m) initially at rest as shown in Figure 9.5. It begins to slide off down and at some location the apple loses contact with the surface. What is the apple's speed when it loses contact?

We will pretend the apple is very small and smooth and slides rather than rolls. Remember we are still dealing with particle kinetics so size and shape are ignored. The apple will begin to

FBD   IBD

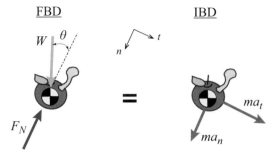

Figure 9.6: FBD/IBD pair of Example 9.2.

leave the surface when the normal force between it and surface is zero. We use an FBD/IBD pair in path coordinates with the apple at an arbitrary angle to setup a relationship (Figure 9.6):

$$\swarrow \sum F_n = ma_n$$

$$- F_N + mg \cos \theta = m \frac{v^2}{r}$$

$$F_N = m \left(g \cos \theta - \frac{v^2}{r} \right).$$

It leaves the surface when $F_N = 0$. Mass cancels and we have a relationship between velocity and angle:

$$v^2 = gr \cos \theta.$$

Apply Work-Energy

$$KE_1 + PE_1 + U_{1 \to 2} = KE_2 + PE_2.$$

The datum is chosen just as the apples leaves the surface

$$y_1 = r (1 - \cos \theta)$$

$KE_1 = 0$

$PE_1 = mgy_1 = mgr (1 - \cos \theta)$

$U_{1 \to 2} = 0$ No external energy added or lost.

$KE_2 = \frac{1}{2}mv^2 = \frac{1}{2}mgr \cos \theta$

$PE_2 = 0$

$$(0) + mgr\,(1 - \cos\theta) + (0) = \frac{1}{2}mgr\cos\theta + (0) \quad \text{(Mass cancels out here too)}$$

$$1 = \frac{3}{2}\cos\theta$$

$$\theta = \cos^{-1}\left(\frac{2}{3}\right) = 48.19°.$$

Enter this angle into the velocity as a function of angle equation:

$$v = \sqrt{gr\cos\theta} = \sqrt{(9.81)\,(0.5)\cos(48.19°)} = 1.808 \text{ m/s}$$

$$\boxed{v = 1.81 \text{ m/s}}.$$

This is a common problem asked in Dynamics courses because it presents a good visualization of motion and challenges a likely misconception that the apple (often a ball) will remain in contact with the surface till it hits the ground. It's interesting that the mass cancels out in both places above.

9.3   POWER

Power is the time rate change of work. In other words, it's how quickly work is performed. We'll use a stylized capitol letter "P" to distinguish it from other uses of that letter in problems (lowercase "p" is often used for pressure, and some books use uppercase "P" for forces). We can write the definition of power in a number of ways:

$$\boxed{\mathbb{P} = \frac{\Delta\,Work}{\Delta\,time} = \frac{dU}{dt} = \frac{d}{dt}\left(\overline{\mathbf{F}}\cdot\overline{\mathbf{r}}\right) = \overline{\mathbf{F}}\cdot\overline{\mathbf{v}}}.$$

The latter form is quite useful in particles. We'll see another useful form when we cover Work-Energy for rigid bodies that relates power to torque and rpm. But for the current topic, force times velocity is a useful concept. It's important to note that power is a scalar (has no direction) measure of an instantaneous state. The dot product in the above equation also reminds us that only the force in line with the velocity matters.

The units of power in SI are: $1 \text{ Watt} = 0.001 \text{ kW} = 1 \text{ J/s} = 1 \text{ Nm/s}$.

The units of power in U.S. Customary are: $1 \text{ horsepower} = 1 \text{ hp} = 550 \text{ ft}\cdot\text{lb/s}$.

The conversion rate between SI and U.S. Customary is: $1 \text{ hp} = 746 \text{ W} = 0.746 \text{ kW}$.

It's often useful to have a frame of reference so we can compare the results in problems asking for power to something we know roughly the size. Here is a brief list of some typical power ratings (maximum) examples gathered from internet searches to keep in mind:

- Weed Wacker: 1.5 hp = 1.12 kW

- Walk Behind Lawnmower: 5 hp = 3.73 kW

- One Horse: 15 hp = 11.19 kW (Figure 9.7)

- Riding Lawnmower: 25 hp = 18.7 kW

- Economy Car: 150 hp = 112 kW

- Sport Utility Vehicle: 250 hp = 187 kW

- High Performance Muscle Car: 450 hp = 336 kW

- Formula One Race Car: 1,000 hp = 746 kW

- Locomotive Engine: 7,000 hp = 5,222 kW

- Jet Engine Gas Turbine: 30,000 hp = 22,400 kW

- Air Craft Carrier: 250,000 hp = 187,000 kW = 187 MW

These are all power ratings which means the maximum power each can achieve. The actual used power for standard operation is usually much less than this value and may never come close to this maximum except for brief periods when the engine is pushed such as acceleration or on a dynamometer by the manufacturer.

Example 9.3
An SUV is rated at 25 mpg (miles per gallon) during highway operation. Make some assumptions, use information from other engineering topics, and approximate the power and overall efficiency.

We assume the speed is

$$v = 65 \text{ mph} = \frac{(65 \text{ mi/hr}) (5280 \text{ ft/mi})}{(3600 \text{ s/hr})} = 95.33 \text{ ft/s}.$$

We'll ignore the road resistance and only look at the wind resistance due to aerodynamic drag. Drag force is often covered in Fluid Mechanics class and can be found from:

$$F_D = \frac{1}{2} C_D \rho v^2 A.$$

The average drag coefficient for an SUV is $C_D = 0.4$ (based on internet search results), where ρ = air density and A = front area.

Density of air at 60°F and standard barometric pressure is $\rho = 0.00238$ slugs/ft^3.

Figure 9.7: Newtdog with one horsepower (actually 14.9 hp) (© E. Diehl).

The front view dimensions of a mid-size SUV are approximately 4.5 ft from bottom of chassis to top of roof and 5.5 ft wide.

The front area (projected area facing direction movement) is therefore $A = (4.5)(5.5) = 24.75$ ft^2.

The air resistance/drag is therefore, approximately:

$$F_D = \frac{1}{2}C_D \rho v^2 A = \frac{1}{2}(0.4)\left(0.00238 \text{ slugs/ft}^3\right)(95.33 \text{ ft/s})^2 \left(24.75 \text{ ft}^2\right)$$
$$= 107.1 \text{ slugs} \cdot \text{ft/s}^2 = 107.1 \text{ lb.}$$

The power is:

$$\mathbb{P} = F_D v = \frac{(107.1 \text{ lb})(95.33 \text{ ft/s})}{\left(550 \frac{\text{ft·lb/s}}{\text{hp}}\right)} = 18.56 \text{ hp}$$

$$\boxed{\mathbb{P} = 18.6 \text{ hp}}.$$

This may seem very small compared to a typical SUV engine rating (250 hp), but remember this is just cruising on the highway not accelerating or pulling a heavy load up a slope where more instantaneous power is required. Also note that this is different than engine power because it's "delivered power." In the efficiency equation, this power is What You Want (WYW).

To find What You Paid For (WYPF) we use the mileage rating to find the energy rate of the fuel.

The energy in gasoline is measured in "heating value" using British Thermal Units (BTU) in U.S. Customary Units. The typical heating value of gasoline is 114,000 BTU/gal. The conversion into equivalent mechanical energy is 1 BTU = 778 ft · lb.

At 65 mph and 25 mpg, the fuel flow is:

$$\frac{(65 \text{ mi/hr})}{(25 \text{ mi/gal})} = 2.600 \text{ gal/hr}.$$

The power available from this fuel flow rate is:

$$\frac{(2.600 \text{ gal/hr}) \, (114{,}000 \text{ BTU/gal}) \, (778 \text{ ft} \cdot \text{lb/BTU})}{(3600 \text{ s/hr}) \left(550 \, \frac{\text{ft} \cdot \text{lb/s}}{\text{hp}}\right)} = 116.5 \text{ hp}.$$

The overall efficiency is therefore:

$$\eta_{ovr} = \frac{WYW}{WYPF} = \frac{(18.56)}{(116.5)} = 15.94\%$$

$$\boxed{\eta_{ovr} = 15.9\%} \,.$$

This is remarkably low, but not unexpected because first the mileage rating includes some speed changes on the highway, so the instantaneous fuel rate during constant speed is likely much lower. Second, there are energy efficiencies for each stage of energy conversion and transmission. Each efficiency can also be thought of as a loss, or an "inefficiency." For example, combustion efficiency ($\sim$ 95%) converting chemical potential energy of fuel into heat in the engine, thermodynamic efficiency ($\sim$ 60%) as you'll cover in thermodynamics idealized as the Otto cycle in a four-stroke engine, mechanical efficiency ($\sim$ 75%) which is friction losses and internal gas flow losses, and the drive train efficiency ($\sim$ 65%) from the transmission and power to wheels losses. There are also auxiliaries like pumps, fans, and air conditioning that use energy and can therefore be treated as losses ($\sim$ 95%). We could estimate the overall efficiency by multiplying all of these:

$$\eta_{est} = \eta_{comb} \eta_{therm} \eta_{mech} \eta_{trans} \eta_{aux} = (0.95) \, (0.60) \, (0.75) \, (0.65) \, (0.95) = \sim 26\%.$$

This example is just an exercise for concept demonstration purposes and gets us into the ballpark quantifying efficiency for a real life application.

Example 9.4

The make-shift horse-powered elevator ($m = 90$ kg) in Figure 9.8 using rope and pulleys is used to lift a basket of apples ($m = 25$ kg). The elevator is to go up 3 floors (9 m) in 6 s with constant

Figure 9.8: Example 9.4 (© E. Diehl).

acceleration till the midway point and then constant deceleration until it reaches the top and stops. Determine the peak power (in horsepower) and the horse's efficiency if she requires a bale of hay after performing this lift 200 times.

Assume: 1 hay bale = 20 kg, alfalfa hay estimated at 900 cal*/lb.

We use kinematics to determine the accelerations and peak velocity of the elevator. We'll find the acceleration to the midway point in half the time and assume the deceleration is the same except negative:

$$y = y_0 + v_0 t + \frac{1}{2}at^2.$$

$$(4.5) = (0) + (0)(3) + \frac{1}{2}a(3)^2$$

$$a = 1.000 \text{ m/s}^2.$$

The peak velocity at the midway point is:

$$v = v_0 + at = (0) + (1.000)(3) = 3.000 \text{ m/s}.$$

For the constraint relationships between the elevator and horse, (Figure 9.9) we cut the focus on a point on the rope we'll call "B" (we can ignore the rest of the rope since it's length is constant):

Cable length: $4S_A + S_B = l$

Time derivative: $\quad v_B = |4v_A|$

2nd derivative: $\quad a_B = |4a_A|$

Just before the midway point, the elevator is still accelerating, so this will be the time of the maximum cable tension and maximum velocity. We draw an FBD/IBD pair of the elevator at that instant (Figure 9.10):

$$\uparrow \sum F_y = 0$$

$$4F_T - mg = ma$$

$$F_T = \frac{1}{4}m\,(g + a)$$

$$F_T = \frac{1}{4}\,(90 + 25)\,(9.81 + 1) = 621.6 \text{ N}.$$

This is also force of tension of the rope the horse pulls. The peak velocity of horse is

$$v_B = |4v_A| = 4\,(3.000) = 12.00 \text{ m/s}.$$

The peak power is:

$$\mathbb{P} = F_D v = (621.6)\,(12.00) = 7{,}459 \text{ N} \cdot \text{m/s} = 7.459 \text{ kW} = 9.999 \text{ hp}$$

$$\boxed{\mathbb{P} = 10.0 \text{ hp}}.$$

To find the energy used we apply Work-Energy:

$$KE_1 + PE_1 + U_{1\to 2} = KE_2 + PE_2$$

$$KE_1 = 0$$

$$PE_1 = 0$$

$$U_{1\to 2} = ?$$

$$KE_2 = 0$$

$$PE_2 = mgy_2 = (90 + 25)\,(9.81)\,(9) = 10{,}150 \text{ Nm}$$

$$(0) + (0) + U_{1\to 2} = (0) + (10{,}150)$$

$$U_{1\to 2} = 10{,}150 \text{ Nm}.$$

The total energy if this lift is performed 200 times:

$$U_{total} = (200)\,(10{,}150) = 2{,}030{,}000 \text{ Nm}.$$

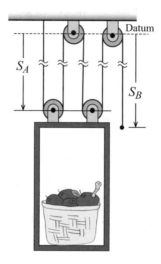

Figure 9.9: Example 9.4 dependent motion relations (© E. Diehl).

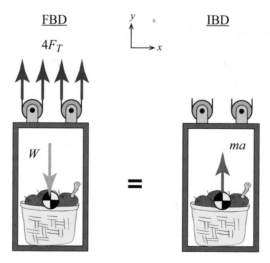

Figure 9.10: FBD/IBD pair of Example 9.4 (© E. Diehl).

To find the energy consumed by the horse we note that "calories" when describing nutrition is different than as applied to thermodynamics. A conversion factor is:

$$1 \ \textit{nutritional Calorie} = 1000 \ \textit{thermodynamic calorie} = 4{,}187 \ \text{Nm}$$

$$U_{hay \ bale} = (20 \ \text{kg}) \ (2.205 \ \text{lb/kg}) \ (900 \ \text{nu cal/lb}) \ (4{,}187 \ \text{Nm/nu cal}) = 16{,}620{,}00 \ \text{Nm}.$$

The horse's efficiency is therefore:

$$\eta_{horse} = \frac{WYW}{WYPF} = \frac{U_{total}}{U_{hay \ bale}} = \frac{(2{,}030{,}000)}{(16{,}620{,}00)} = 12.22\%$$

$$\boxed{\eta_{horse} = 12.2\%} \ .$$

REFERENCES:

Gerhart, P. M., Gerhart, A. L., and Hochstein, J. I. *Munson, Young and Okiishi's Fundamentals of Fluid Mechanics*. John Wiley & Sons, 2016.

https://en.wikipedia.org/wiki/Horsepower

https://www.fueleconomy.gov/feg/atv.shtml

For a similar analysis see:
https://dothemath.ucsd.edu/2011/07/100-mpg-on-gasoline/

Book 1 - Class 10

https://www.youtube.com/watch?v=vdcdNrDtmME

CLASS 10

Impulse-Momentum Method

B.L.U.F. (Bottom Line Up Front)

- Linear Impulse: $\overrightarrow{\textbf{IMP}} = \int_{t_1}^{t_2} \overrightarrow{\textbf{F}}\, dt.$

- Linear Momentum: $\overrightarrow{\textbf{L}} = m\overrightarrow{\textbf{v}}.$

- Impulse-Momentum: $m\overrightarrow{\textbf{v}}_1 + \int_{t_1}^{t_2} \overrightarrow{\textbf{F}}\, dt = m\overrightarrow{\textbf{v}}_2.$

- Impulse-Momentum involves vectors while Work-Energy uses scalars.

- Impulse-Momentum with kinematics can be used in lieu of N2L or Work-Energy in some situations.

10.1   LINEAR IMPULSE-MOMENTUM

In Class 6 we wrote N2L in terms of the time rate change of linear momentum. Momentum is mass × velocity ($\overrightarrow{\textbf{L}} = m\overrightarrow{\textbf{v}}$) and is described by Newton's First Law: an object in motion tends to stay in motion. A large object, such as the large wagon full of apples in Figure 10.1 starting from rest, is tough to get moving as Newtdog starts to push. After some time pushing, it starts to gather speed. And once it's moving, it's difficult to stop. This is linear momentum.

Now recall from N2L: $\sum \overrightarrow{\textbf{F}} = \frac{d}{dt}(\overrightarrow{\textbf{L}}) = \frac{d}{dt}(m\overrightarrow{\textbf{v}}).$

If we rearrange $\sum \overrightarrow{\textbf{F}} = \frac{d}{dt}(m\overrightarrow{\textbf{v}})$ we get $\sum \overrightarrow{\textbf{F}}\, dt = m\, d\overrightarrow{\textbf{v}}$ if mass remains constant. We can integrate this between two points:

$$\int_{t_1}^{t_2} \overrightarrow{\textbf{F}}\, dt = m\overrightarrow{\textbf{v}}_2 - m\overrightarrow{\textbf{v}}_1.$$

Rearranging we get the main Linear Impulse-Momentum equation:

$$\boxed{m\overrightarrow{\textbf{v}}_1 + \int_{t_1}^{t_2} \overrightarrow{\textbf{F}}\, dt = m\overrightarrow{\textbf{v}}_2}.$$

This can also be rewritten as:

$$\overrightarrow{\textbf{L}}_1 + \overrightarrow{\textbf{IMP}}_{1\to2} = \overrightarrow{\textbf{L}}_2,$$

Figure 10.1: Newtdog uses impulse to create linear momentum (© E. Diehl).

or

$$Momentum\ Before + Applied\ Impulse = Momentum\ After.$$

In the moving apple wagon above, Newtdog is imparting the impulse on the wagon with his pushing force multiplied by the time he pushes. The same would be true to stop the wagon, but he'd need to push it in the other direction. This reminds us that Impulse-Momentum depends on direction because it is a vector equation.

To visualize this process and to keep track of what's going on when solving problems, we use an Impulse-Momentum diagram such as Figure 10.2. This diagram corresponds to elements of the linear Impulse-Momentum equation broken down into components:

x-dir:

$$mv_{x1} + \int_{t_1}^{t_2} F_x dt = mv_{x2}$$

y-dir:

$$mv_{y1} + \int_{t_1}^{t_2} F_y dt = mv_{y2}.$$

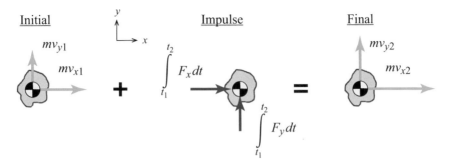

Figure 10.2: Impulse-Momentum diagram.

The units of both momentum and impulse are:

$$m\vec{v} = (\text{kg})\,(\text{m/s}) = \text{N}\cdot\text{s} \quad \text{or} \quad m\vec{v} = (\text{slug})\,(\text{ft/s}) = \text{lb}\cdot\text{s}$$

(recalling that 1 lb = 1 slugft/s^2).

$$\vec{F}\,dt = (\text{N})\,(\text{s}) = \text{N}\cdot\text{s} \quad \text{or} \quad \vec{F}\,dt = (\text{lb})\,(\text{s}) = \text{lb}\cdot\text{s}.$$

10.2   IMPULSE WITH LARGE FORCE AND SMALL TIME

Impulse is often a large force applied for a very short duration. The time can be too short to accurately measure even with using a high speed camera. Similarly the impulsive force during a collision or impact can be so large it is difficult to measure. When this is the case, we'll often lump the force and time together and consider impulse a parameter ($\overrightarrow{\textbf{IMP}}$) we're either given or solving for:

$$\int_{t_1}^{t_2} \vec{F}\,dt = \overrightarrow{\textbf{IMP}}_{1\rightarrow2}.$$

In Figure 10.3 we show Newtdog imparting impulse onto Wormy's apple. So we can rewrite the Impulse-Momentum equation as: $m\vec{v}_1 + \overrightarrow{\textbf{IMP}}_{1\rightarrow2} = m\vec{v}_2$.

For short duration impulses, the weight is often neglected since its magnitude is relatively small compared to the other impulsive forces and multiplied by a very small time duration as shown in Figure 10.4.

10.3   CONSTANT FORCE OR AVERAGE FORCE IMPULSE

If the impulsive force is constant or if we use choose to the average force, we can remove the integration: $\int_{t_1}^{t_2} \vec{F}\,dt = \sum \vec{F}_{AVG}\Delta t$.

Integration is, of course, the area under the curve of force vs. time (see Figure 10.5), so we can divide $\overrightarrow{\textbf{IMP}}$ by Δt to obtain an equivalent average force to use for the Impulse-Momentum

Figure 10.3: Newtdog imparts impulse onto Wormy's apple (© E. Diehl).

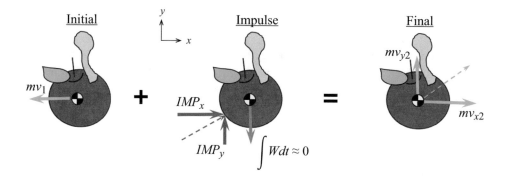

Figure 10.4: Impulse-Momentum diagram showing weight as negligible (© E. Diehl).

equation:

$$m \vec{\mathbf{v}}_1 + \sum \vec{\mathbf{F}}_{AVG} \Delta t = m \vec{\mathbf{v}}_2.$$

Examples of this situation are the impulse thrust of space craft (the real thing, not just in science fiction) where a timed release of gases propels the craft in a direction, therefore changing its momentum.

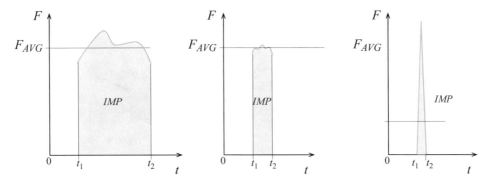

Figure 10.5: Impulse is area under force vs. time curve.

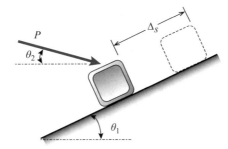

Figure 10.6: Example 10.1, repeat of Figure 6.6.

We'll use constant force impulse for several examples, including two examples that previously used both N2L and Work-Energy in order to demonstrate that Impulse-Momentum is a related Kinetics method.

Example 10.1   (This is a repeat of Examples 6.1 and 8.1)
A constant $P = 50$ lb force is applied to a box weighing 25 lb, starting from rest, and positioned on a $\theta_1 = 25°$ inclined surface with $\mu_s = 0.25$ and $\mu_k = 0.2$ static and kinetic coefficients of friction, respectively (Figure 10.6). The force is applied to box $\theta_2 = 15°$ from horizontal. Determine the distance up the slope box travels (Δs) when it reaches a speed of $v = 3$ ft/s.

Recall from Examples 6.1 and 8.1 that we need to do an FBD/IBD pair in order to find the normal force and therefore the friction force (Figure 10.7):

$$\nwarrow \sum F_{y'} = ma_{y'} = 0$$

$$- P \sin (\theta_1 + \theta_2) - W \cos \theta_1 + F_N = 0$$

$$F_N = (50) \sin (40°) + (25) \cos (25°) = 54.80 \text{ lb}$$

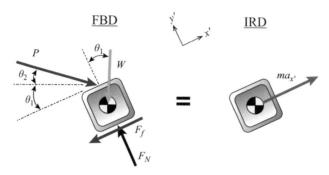

Figure 10.7: FBD/IBD of Example 10.1, repeat of Figure 6.7.

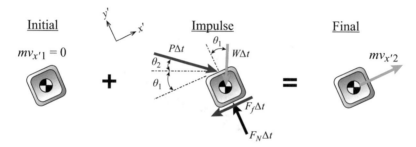

Figure 10.8: Impulse-Momentum diagram of Example 10.1.

$$F_f = \mu_k F_N = (0.20)(54.80) = 10.96 \text{ lb}.$$

We draw an Impulse-Momentum diagram (Figure 10.8):

x'-dir:

$$mv_{x'1} + \int_{t_1}^{t_2} F_{x'}dt = mv_{x'2}$$

$$mv_{x'1} + P\Delta t \cos(\theta_1 + \theta_2) - W\Delta t \sin\theta_1 - F_f \Delta t = mv_{x'2}$$

$$(0) + (50)\Delta t \cos(25° + 15°) - (25)\Delta t \sin(25°) - (10.96)\Delta t = \left(\frac{25}{32.2}\right)(3)$$

$$\Delta t = 0.1388 \text{ s}.$$

From kinematics we can find the acceleration (which is constant because the forces are constant):

$$v = v_0 + at$$

$$(3) = (0) + a(0.1388)$$

$$a = 21.61 \text{ ft/s}^2.$$

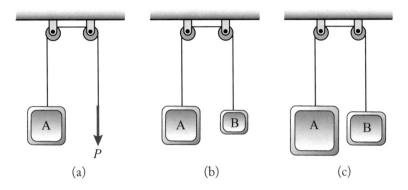

Figure 10.9: Example 10.2, repeat of Figure 6.11.

Which we can use to find the distance also from kinematics:

$$v^2 = v_0^2 + 2a\Delta s$$

$$\Delta s = \frac{v^2 - v_0^2}{2a} = \frac{(3)^2 - 0^2}{2\,(21.61)} = 0.2083 \text{ ft} = 2.499 \text{ in}$$

$$\boxed{\Delta s = 2.50 \text{ in}}.$$

We see that we get the same results from Examples 5.1 and 8.1.

Example 10.2   (This is a repeat of Examples 6.3 and 8.2)
The three setups shown begin at rest. In setup (a) the force ($P = 50$ lb) is applied to the cable attached to block A ($W_A = 75$ lb) (Figure 10.9). Setup (b) has the same block A and is connected to block B ($W_B = 50$ lb). Setup (c) has larger blocks with the same difference between them ($W_A = 175$ lb and $W_B = 150$ lb). The pulleys are assumed to be massless and frictionless. Determine for each setup the acceleration of block A.

This example demonstrates it's possible, however a bit more work, to use Impulse-Momentum to solve the same problem as Examples 6.3 and 8.2 which use N2L and Work-Energy, respectively.

Part (a)
Using the Impulse-Momentum diagram of Part (a) (drawn with F_T so it can be used in Parts (b) and (c)) (Figure 10.10):

$$m v_{y1} + \int_{t_1}^{t_2} F_y dt = m v_{y2}$$

$$(0) + F_T \Delta t - W_A \Delta t = -m_A v_{Ay2} \ \textcircled{1}$$

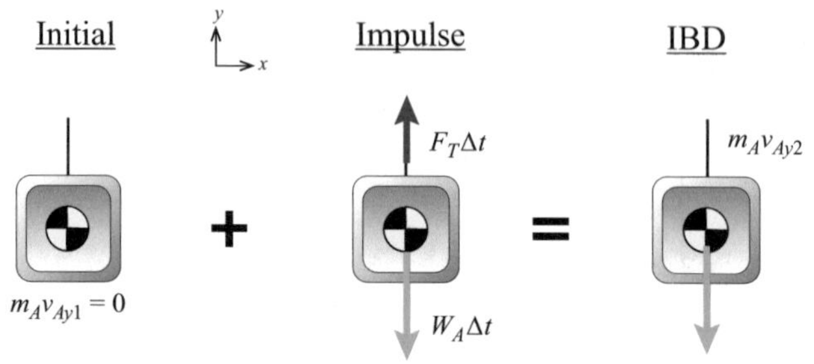

Figure 10.10: Example 10.2 Part (a) Impulse-Momentum diagram.

$$F_T = P = 50 \, \text{lb}$$

$$(0) + (50) \, \Delta t - (75) \, \Delta t = - \left(\frac{75}{32.2} \right) v_{Ay2}$$

$$(25) \, \Delta t = (2.329) \, v_{Ay2}.$$

We are left with two unknowns, neither of which are acceleration. But since we know accelera-tion must be constant, we can choose an arbitrary increment of time and find the velocity. Let's use $\Delta t = 1$ s:

$$(25) (1) = (2.329) \, v_{Ay2}$$

$$v_{Ay2} = 10.73 \, \text{ft/s}.$$

Acceleration can be found from:

$$v = v_0 + at$$

$$(10.73) = (0) + a (1)$$

$$a = 10.73 \, \text{ft/s}^2$$

(a) $\boxed{\vec{a}_A = 10.73 \, \text{ft/s}^2 \downarrow}$ This matches the answers in Examples 6.3 and 8.2.

Part (b)

We can re-use the Impulse-Momentum diagram from Part (a) for block A. We need a separate Impulse-Momentum diagram for block B (Figure 10.11). This is an important aspect of this problem to point out: we CANNOT omit any external forces from the impulses so we CAN-NOT do this problem with one diagram since we'd need to include the pulley whose reaction force we don't know. Instead, we use separate diagrams linked with the tension of the rope:

$$mv_{y1} + \int_{t_1}^{t_2} F_y dt = mv_{y2}$$

$$(0) + F_T \Delta t - W_B \Delta t = mv_{By2} \, \textcircled{2} .$$

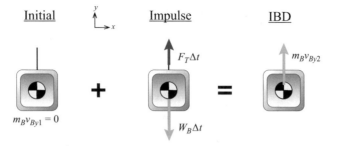

Figure 10.11: Example 10.2 Part (b) Impulse-Momentum diagram.

We can use equations ① and ②, and the dependent relation $v_{Ay2} = |v_{By2}|$. From equation ①:

$$F_T \Delta t - W_A \Delta t = -m_A v_{Ay2} \qquad F_T = W_A - \frac{m_A v_{Ay2}}{\Delta t} \quad ③ .$$

From equation ②:

$$F_T \Delta t - W_B \Delta t = m_B v_{By2} \qquad F_T = W_B + \frac{m_B v_{By2}}{\Delta t} \quad ④ .$$

Equating equations ③ and ④ and replacing $v_{Ay2} = |v_{By2}|$:

$$W_A - \frac{m_A v_{By2}}{\Delta t} = W_B + \frac{m_B v_{By2}}{\Delta t}$$

$$W_A - W_B = \frac{m_A v_{By2}}{\Delta t} + \frac{m_B v_{By2}}{\Delta t}$$

$$(W_A - W_B) \Delta t = (m_A + m_B) v_{By2}$$

$$v_{By2} = \frac{(W_A - W_B) \Delta t}{(m_A + m_B)} \quad ⑤ .$$

Inputting values and the assumption of $\Delta t = 1$s we get:

$$v_{By2} = \frac{((75) - (50)) (1)}{\left(\left(\frac{75}{32.2}\right) + \left(\frac{50}{32.2}\right)\right)} = 6.440 \text{ ft/s}.$$

Acceleration from:

$$v = v_0 + at$$

$$(6.440) = (0) + a (1)$$

$$a = 6.440 \text{ ft/s}^2.$$

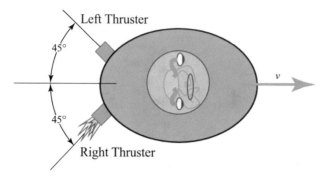

Figure 10.12: Example 10.3 thrust force vs. time.

(b) $\boxed{\vec{a}_A = 6.44 \text{ ft/s}^2 \downarrow}$ This too matches the answers in Examples 6.3 and 8.2.

Part (c)

We can re-use the results of part (b) and input the new values of the weights into equation ⑤:

$$v_{By2} = \frac{(W_A - W_B)\,\Delta t}{(m_A + m_B)}$$

$$v_{By2} = \frac{((175) - (150))\,(1)}{\left(\left(\frac{175}{32.2}\right) + \left(\frac{150}{32.2}\right)\right)} = 2.477 \text{ ft/s}.$$

Acceleration from:

$$v = v_0 + at$$

$$(2.477) = (0) + a\,(1)$$

$$a = 2.477 \text{ ft/s}^2.$$

(c) $\boxed{\vec{a}_A = 2.48 \text{ ft/s}^2 \downarrow}$ This also matches the answers in Examples 6.3 and 8.2.

Example 10.3

A spaceship ($m = 1000$ kg) is traveling in a straight line at 2000 m/s, when the pilot applies the right thruster (arranged to provide both forward and side thrust as shown in Figure 10.13) for $t = 10$ s. The thruster ramps up the applied force as shown in the graph, peaking at 200 kN (Figure 10.12). Determine the *velocity* after the thrust has been applied.

Apply impulse and linear momentum

$$m\vec{v}_1 + \int_{t_1}^{t_2} \vec{F}\,dt = m\vec{v}_2.$$

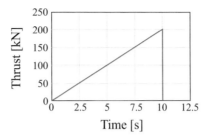

Figure 10.13: Example 10.3 spacecraft.

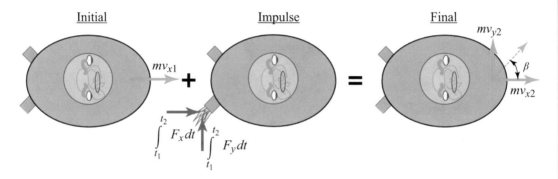

Figure 10.14: Impulse-Momentum diagram of Example 10.3.

Noting that force changes with time, we can't just use the force times Δt, but we do know the integral of this curve will be the area under the triangle, so instead of writing an equation for the sloped line and integrating, we can just say:

$$\int_{t_1}^{t_2} \vec{\mathbf{F}}\, dt = \frac{1}{2} F_{\max} \Delta t = \frac{1}{2} (200{,}000)(10) = 1{,}000{,}000 \text{ N} \cdot \text{s} \nearrow 45°.$$

The impulse is still a vector, so we need to do this in components (Figure 10.14):

$$\int_{t_1}^{t_2} F_x\, dt = \cos 45° \int_{t_1}^{t_2} \vec{\mathbf{F}}\, dt = \cos 45° (1{,}000{,}000) = 707{,}100 \text{ N} \cdot \text{s} \rightarrow$$

$$\int_{t_1}^{t_2} F_y\, dt = \sin 45° \int_{t_1}^{t_2} \vec{\mathbf{F}}\, dt = \sin 45° (1{,}000{,}000) = 707{,}100 \text{ N} \cdot \text{s} \uparrow.$$

x-dir:

$$mv_{x1} + \int_{t_1}^{t_2} F_x\, dt = mv_{x2} \qquad (1000)(2000) + (707{,}100) = (1000)\, v_{x2}$$

$$v_{x2} = 2{,}707 \text{ m/s}.$$

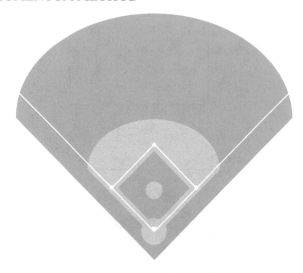

Figure 10.15: Example 10.4.

y-dir:

$$mv_{y1} + \int_{t_1}^{t_2} F_y\, dt = mv_{y2} \qquad (1000)\,(0) + (707{,}100) = (1000)\,v_{y2}$$

$$v_{y2} = 707.1 \text{ m/s}$$

$$|v_2| = \sqrt{v_{x2}{}^2 + v_{y2}{}^2} = \sqrt{(2{,}707)^2 + (707.10)^2} = 2{,}798 \text{ m/s} \quad \beta = \tan^{-1}\left(\frac{707.10}{2{,}707}\right) = 14.64°$$

$$\boxed{\vec{v}_2 = 2{,}800 \text{ m/s} \nearrow 14.6°} \; .$$

Example 10.4

A baseball is thrown at 95 mph, and the batter hits it over an outfield fence where it lands 500 ft away (on the ground at the same level as the batter). The baseball reaches a zenith of 90 ft (Figure 10.15). A high-speed camera shows that the ball remains in contact with the bat for $\Delta t = 0.015$ s. Estimate the average force and angle that the bat exerts onto the baseball during impact. (Note: a typical baseball weighs 5 oz.)

We can use projectile motion to find the velocity (speed and angle) of the ball as it leaves the bat. We can find the initial vertical velocity from the peak distance:

$$v_y{}^2 = \left(v_y\right)_0^2 - 2g\Delta y$$

$$(0)^2 = \left(v_y\right)_0^2 - 2\,(32.2)\,(90)$$

$$\left(v_y\right)_0 = 76.13 \text{ ft/s.}$$

We can find the time of flight by the time to reach this peak and doubling it:

$$v_y = \left(v_y\right)_0 - gt$$

$$(0) = (76.13) - (32.2)\, t$$

$$t = 2.364 \text{ s.}$$

The ball lands at $2t = 4.729$ s.

The initial horizontal velocity is found from:

$$x = x_0 + \left(v_x\right)_0 t$$

$$(500) = (0) + \left(v_x\right)_0 (4.729)$$

$$\left(v_x\right)_0 = 105.7 \text{ ft/s.}$$

The angle of the initial velocity is $\theta = \tan^{-1} \frac{\left(v_y\right)_0}{\left(v_x\right)_0} = \tan^{-1} \frac{(76.13)}{(105.7)} = 35.76°$. This is for information since the bat's impulse angle won't match this value, which is an interesting result and dispels a common misconception that you must hit the baseball in the direction you want it to go. We'll switch the designation to help with the rest of the problem, so $v_{x2} = 105.7$ ft/s and $v_{y2} = 76.13$ ft/s.

The ball's speed is

$$v_{x1} = \frac{(95 \text{ mi/hr}) (5280 \text{ ft/mi})}{(3600 \text{ s/hr})} = 139.3 \text{ ft/s.}$$

Mass of ball:

$$m = \frac{(5) / (16)}{(32.2)} = 9.705 \cdot 10^{-3} \text{ slugs.}$$

Using the Impulse-Momentum diagram (Figure 10.16)

x-dir:

$$-mv_{x1} + F_x \Delta t = mv_{x2}$$

$$-\left(9.705 \cdot 10^{-3}\right)(139.3) + F_x (0.015) = \left(9.705 \cdot 10^{-3}\right)(105.7)$$

$$F_x = 158.5 \text{ lb} \rightarrow$$

y-dir:

$$mv_{y1} + F_y \Delta t = mv_{y2}$$

$$(0) + F_y (0.015) = \left(9.705 \cdot 10^{-3}\right)(76.13)$$

$$F_y = 49.26 \text{ lb} \uparrow$$

$$|\vec{\mathbf{F}}| = \sqrt{(158.5)^2 + (49.26)^2} = 166.0 \text{ lb} \quad \theta \tan^{-1}\left(\frac{49.26}{158.5}\right) = 17.26°$$

$$\boxed{\vec{\mathbf{F}} = 166.0 \text{ lb} \nearrow 17.26°}.$$

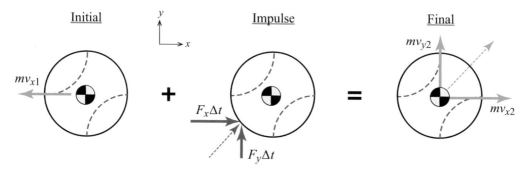

Figure 10.16: Example 10.4 Impulse-Momentum diagram.

Table 10.1: Summary of particle kinetics methods

Method	Equation	Scalar or Vector	Key Parameters
N2L	$\sum \vec{\mathbf{F}} = m\vec{\mathbf{a}}$	Vector	Force and Acceleration
Work-Energy	$KE_1 + PE_1 + U_{1\to2} = KE_2 + PE_2$	Scalar	Force, Velocity, and Displacement
Linear Impulse-Momentum	$m\vec{\mathbf{v}}_1 + \int_{t_1}^{t_2} \vec{\mathbf{F}}\, dt = m\vec{\mathbf{v}}_2$	Vector	Force, Velocity, and Time

10.4   COMPARISON OF PARTICLE KINETICS METHODS

From Examples 10.1 and 10.2 we have confirmed it is possible to use any of the three methods we've covered in particle kinetics to solve some problems. There are many other scenarios, where it is either inconvenient or even impossible to solve using multiple methods (as we'll see in the next two classes on impact collisions). One such situation is the pendulum problem in Examples 7.1 and 8.3 where the linear impulse method can't be used to solve it. Angular Impulse-Momentum (covered in more detail Rigid Body Kinetics), however, can be used to solve the pendulum problem.

Table 10.1 is a summary of the particle kinetics methods and the key parameters of each. Kinematics can, in some cases, link these methods, but some problems are much more difficult to solve this way. Note that Work-Energy removes the direction since it is a scalar method which can have advantages and disadvantages.

10.5 ANGULAR IMPULSE-MOMENTUM

When particles rotate about an axis they also possess angular momentum defined as $\vec{\mathbf{H}}_O = \vec{\mathbf{r}} \times m\vec{\mathbf{v}}$, where the position vector $\vec{\mathbf{r}}$ is the particle distance from the reference axis O. Angular impulse occurs from a moment applied for a short duration or can be thought of as linear impulse with a moment arm. This is written as

$$\overrightarrow{\mathbf{ANGIMP}}_{1\to2} = \int_1^2 \vec{\mathbf{M}}_O dt = \int_{t_1}^{t_2} (\vec{\mathbf{r}} \times \vec{\mathbf{F}})\, dt.$$

The Angular Impulse-Momentum equation is written:

$$(\vec{\mathbf{H}}_O)_1 + \overrightarrow{\mathbf{ANGIMP}}_{1\to2} = (\vec{\mathbf{H}}_O)_2.$$

For one particle:

$$\vec{\mathbf{r}} \times m\vec{\mathbf{v}}_1 + \int_1^2 \vec{\mathbf{M}}_O dt = \vec{\mathbf{r}} \times m\vec{\mathbf{v}}_2.$$

For multiple particles:

$$\sum \vec{\mathbf{r}} \times m\vec{\mathbf{v}}_1 + \sum \int_1^2 \vec{\mathbf{M}}_O dt = \sum \vec{\mathbf{r}} \times m\vec{\mathbf{v}}_2.$$

When there is no external moment applied to a particle rotating about a fixed axis, and therefore no angular impulse, angular momentum is conserved. A system of particles can have a total angular momentum which is conserved if no external influence changes it. Angular momentum of a system of particles is quite similar to angular momentum of rigid bodies and the topic is often included in textbooks as a transition to discussing rigid body angular momentum. A simple example follows, and more discussion of angular momentum is contained in Class 23 (vol. 2) as applied to rigid bodies.

Example 10.5
Two apples are on a stick that is about to be spun as shown in Figure 10.17. Apple A is $m_A = 100$ g and $r_A = 100$ mm from the pivot point, and is Apple B is $m_B = 70$ g and $r_B = 150$ mm from the pivot point.

1. Determine the constant moment, M, that is required to be applied for $\Delta t = 10$ s in order for Apple B to reach $v_B = 100$ m/s if it is applied to the stick starting from rest.

2. If after the moment is removed, Apple A slips to $r_A = 150$ mm while Apple B remains at $r_B = 150$ mm, what is the final speed of Apples A and B?

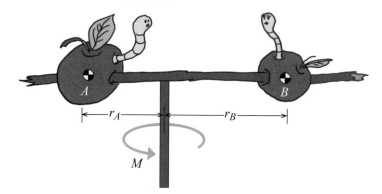

Figure 10.17: Apples on a stick in Example 10.5 (© E. Diehl).

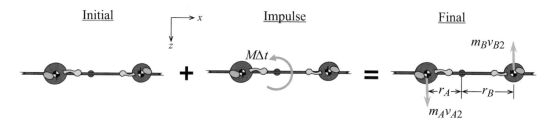

Figure 10.18: Impulse-Momentum diagram of Example 10.5 Part (a) (© E. Diehl).

The velocity of Apple A after the moment is removed is found from kinematics:

$$v_{B2} = r_B\omega_2 \quad \omega_2 = \frac{v_{B2}}{r_B} = \frac{(100)}{(0150)} = 666.7 \text{ rad/s}$$

$$v_{A2} = r_A\omega_2 = (0.100)(666.7) = 66.67 \text{ m/s}.$$

Figure 10.18 shows the Impulse-Momentum diagram in the x–z plane:

$$\sum \vec{r} \times m\vec{v}_1 + \sum \int_1^2 \vec{M}_O dt = \sum \vec{r} \times m\vec{v}_2.$$

Because the moment is constant, the angular impulse integral is simply $\int_1^2 \vec{M}_O dt = M\Delta t$

$$r_A m_A v_{A1} + r_B m_B v_{B1} + M\Delta t = r_A m_A v_{A2} + r_B m_B v_{B2}$$

$$(0.100)(0.100)(0) + (0.150)(0.070)(0) + M(10)$$
$$= (0.100)(0.100)(66.67) + (0.150)(0.070)(100)$$

$$M = 0.1717 \text{ Nm}.$$

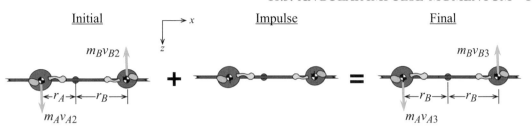

Figure 10.19: Impulse-Momentum diagram of Example 10.5 Part (b) (© E. Diehl).

After the moment is removed, Apple A slips to $r_A = r_B = 150$ mm. Since there is no external impulse, this is a conservation of angular momentum problem. From kinematics we know that the final velocities will be equal when the distances from the pivot point are equal. Figure 10.19 shows the Impulse-Momentum diagram of this scenario.

$$\sum \vec{\mathbf{r}} \times m\vec{\mathbf{v}}_1 = \sum \vec{\mathbf{r}} \times m\vec{\mathbf{v}}_2$$

$$r_A m_A v_{A2} + r_B m_B v_{B2} = r_A m_A v_{A3} + r_B m_B v_{B3}$$

$$(0.100)(0.100)(66.67) + (0.150)(0.070)(100) = (0.150)(0.100)v_{A2} + (0.150)(0.070)v_{B2}$$

$$v_{A2} = v_{B2} = 67.32 \text{ m/s}.$$

Answers:

   (a) $\boxed{M = 0.172 \text{ Nm}}$

   (b) $\boxed{v_{A2} = v_{B2} = 67.3 \text{ m/s}}$

Book 1 - Class 11

https://www.youtube.com/watch?v=96gL_myCHdg

CLASS 11

Direct Impact of Particles and the Conservation of Linear Momentum

B.L.U.F. (Bottom Line Up Front)

- Conservation of Linear Momentum: $\sum m\vec{\mathbf{v}}_1 = \sum m\vec{\mathbf{v}}_2$.

- Impacts/collisions depend on Coefficient of Restitution (CoR): $e = \frac{v_{B2}-v_{A2}}{v_{A1}-v_{B1}}$.

- CoR is the ratio of relative velocities after and (negative) before.

- "Perfectly Plastic" collision: $e = 0$; "Perfectly Elastic" collision: $e = 1$.

- CoR is similar in concept to the efficiency of collisions since energy is not always conserved.

11.1 IMPULSE-MOMENTUM OF MULTIPLE PARTICLES AND CONSERVATION OF LINEAR MOMENTUM

We can apply Impulse-Momentum to multiple particles such as the cue ball and apple in Figure 11.1, so we can write:

$$\sum m\vec{\mathbf{v}}_1 + \int_{t_1}^{t_2} \vec{\mathbf{F}}\,dt = \sum m\vec{\mathbf{v}}_2.$$

Most often the impulses are due to the interaction (collisions) of the particles. When this is the case, we can treat impulsive forces as "internal" to the system (and therefore equal and opposite), so momentum is conserved. The Linear Impulse-Momentum equation reduces to the "Conservation of Linear Momentum":

$$\boxed{\sum m\vec{\mathbf{v}}_1 = \sum m\vec{\mathbf{v}}_2}.$$

As shown in Figure 11.2, the collision impulses are equal and opposite which cancel each other out when both particles are included in the analysis.

Figure 11.1: Newtdog plays billiards with Wormy (© E. Diehl).

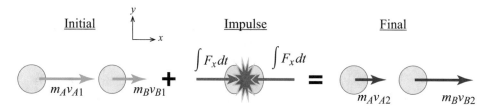

Figure 11.2: Impulse Momentum of a collision demonstrating internal impulses cancel when both particles are included in the analysis.

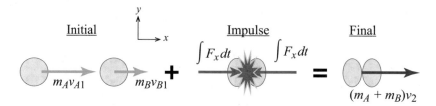

Figure 11.3: Impulse Momentum of a collision demonstrating where particles stick together.

For the first example we will use Conservation of Momentum for the special case where particles stick together, as shown in Figure 11.3.

We will also check the initial and final energy so we can compare the Conservation of Linear Momentum with the Conservation of Energy. We note that it's possible that momentum is conserved while energy is *not* conserved. This is because the collision loses energy and, consequently, an efficiency.

Example 11.1
A car accident at an intersection results in both vehicles stuck together and sliding as one particle, as shown in Figure 11.4. Truck A ($m_A = 3,175$ kg) is traveling at 65 mph in 25° South of East.

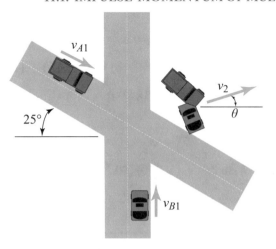

Figure 11.4: Example 11.1.

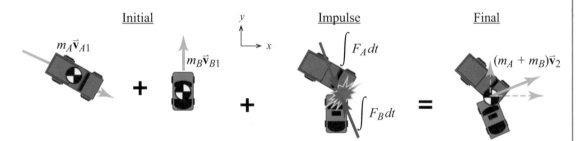

Figure 11.5: Impulse-Momentum diagram of Example 11.1.

Car B ($m_B = 1{,}250$ kg) is traveling at 90 mph North. Determine the velocity the two cars move afterward and the efficiency the collision.

Note that Figure 11.4 shows the final motion as North-East which isn't necessarily the case. It's simply assuming the positive direction.

A bit of housekeeping:

$$v_{A1} = \frac{(65 \text{ mi/hr}) (5280 \text{ ft/mi})}{(3600 \text{ s/hr}) (3.281 \text{ ft/m})} = 29.06 \text{ m/s} \qquad v_{B1} = \frac{(90 \text{ mi/hr}) (5280 \text{ ft/mi})}{(3600 \text{ s/hr}) (3.281 \text{ ft/m})} = 40.23 \text{ m/s}.$$

(Actually don't need to do this as the units would work out anyway.)

The Impulse-Momentum diagram is shown in Figure 11.5.

Applying the Conservation of Momentum

$$\sum m\vec{\mathbf{v}}_1 = \sum m\vec{\mathbf{v}}_2$$

x-dir:

$$m_A v_{A1} \cos 25° + m_B (0) = (m_A + m_B) v_{x2}$$

$$(3,175)(29.06) \cos 25° + (1,250)(0) = ((3,175) + (1,250)) v_{x2}$$

$$v_{x2} = 18.90 \text{ m/s.}$$

y-dir:

$$-m_A v_{A1} \sin 25° + m_B v_{A1} = (m_A + m_B) v_{y2}$$

$$-(3,175)(29.06) \sin 25° + (1,250)(40.23) = ((3,175) + (1,250)) v_{y2}$$

$$v_{y2} = 2.552 \text{ m/s.}$$

$$|\vec{v}_2| = \sqrt{v_{x2}^2 + v_{y2}^2} = \sqrt{(18.90)^2 + (2.552)^2} = 19.07 \text{ m/s}$$

$$\theta = \tan^{-1} \frac{v_{y2}}{v_{x2}} = \tan^{-1} \frac{(2.552)}{(18.90)} = 7.690°.$$

Since we were given the speed in mph, we can infer the answer should be given in the same units:

$$|\vec{v}_2| = \frac{(19.07 \text{ m/s})(3600 \text{ s/hr})(3.281 \text{ ft/m})}{(5280 \text{ ft/mi})} = 42.66 \text{ mph}$$

$$\boxed{\vec{v}_2 = 42.7 \text{ mph} \nearrow 7.69°}.$$

Energy Before:

$$KE_1 = \frac{1}{2} m_A v_{A1}^2 + \frac{1}{2} m_B v_{B1}^2 = \frac{1}{2}(3,175)(29.06)^2 + \frac{1}{2}(1,250)(40.23)^2 = 2,352,000 \text{ Nm.}$$

Energy After:

$$KE_2 = \frac{1}{2}(m_A + m_B) v_2^2 = \frac{1}{2}((3,175) + (1,250))(19.07)^2 = 804,600 \text{ Nm.}$$

The difference between these is energy lost to a variety of places including heat, plastic deformation of metal, and sound.

The efficiency is the resulting energy ("what you want") divided by the input energy ("what you paid for"):

$$\eta = \frac{WYW}{WYPF} = \frac{KE_2}{KE_1} = \frac{(804,600)}{(2,352,000)} = 34.21\% \qquad \boxed{\eta = 34.2\%}.$$

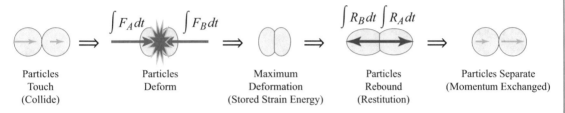

Figure 11.6: Impulse Momentum of a collision demonstrating internal impulses canceling (repeat of Figure 11.2).

Figure 11.7: Stages with impact of two particles.

11.2   DIRECT IMPACT OF PARTICLES AND THE COEFFICIENT OF RESTITUTION

Let's revisit the general case of a two particle collision where particle A catches up with particle B.

The Conservation of Linear Momentum applied to Figure 11.6 (repeat of Figure 11.2) is written as:

x-dir:

$$m_A v_{A1} + m_B v_{B1} = m_A v_{A2} + m_B v_{B2}.$$

We usually know the initial velocities and want to find the final velocities, and therefore have this one equation and two unknowns. We need to know something else to solve this. In the previous example we were told the two stuck together, so we only had one unknown and could solve the problem. So if they don't stick together, what do we do? To find v_{A2} and v_{B2} we need to know something else.

Let's look more closely at the collision process. Figure 11.7 breaks the collision down into five steps.

The particles touch and begin to deform with equal and opposite forces (that are changing with time as the particles compress). The particles reach a point where they've deformed in compression their maximum amount. At this point, these have stored strain energy, similar to springs. The stored energy is released as the particles rebound, a process we call "restitution" that has reaction forces that change with time (but still equal and opposite with the collision). The particles separate after the restitution finishes and linear momentum has been exchanged. This process depends on the way in which the particles deform and rebound.

We showed in Example 11.1 that energy was not conserved but instead lost to a variety of places including heat, plastic deformation of metal, and sound. To address this we introduce a parameter (e) called the "Coefficient of Restitution" (abbreviated here as CoR) defined as:

$$e = \frac{v_{B2} - v_{A2}}{v_{A1} - v_{B1}}.$$

We note that the numerator is the relative velocity after the collision and the denominator is the relative velocity before the collision with the order of the velocity difference reversed. So another way to think of CoR is:

$$e = \frac{Rel\ Velocity\ After}{-Rel\ Velocity\ Before}.$$

We can understand the CoR better by looking at extreme cases.

- Perfectly Plastic Collision: What happens with the particles stick together as in Example 11.1? The relative velocity after the collision is zero. Therefore $e = 0$ when they stick together, a stiution we call "perfectly plastic" because there is no rebound and release of the strain energy.

- Perfectly Elastic Collision: What happens if there is no loss during restitution ($e = 1$)? CoR is also often written as: $v_{B2} - v_{A2} = e\,(v_{A1} - v_{B1})$. So when $e = 1$, $v_{B2} - v_{A2} = v_{A1} - v_{B1}$ which means the relative velocity reverses. This makes sense if you consider the situation where both particles in Figure 11.6 have the same mass and $e = 1$. The velocity of A becomes the velocity of B and vice versa after the collision. We call this "perfectly elastic."

We use the two equations below to find the two unknowns: v_{A2} and v_{B2}.

$$m_A v_{A1} + m_B v_{B1} = m_A v_{A2} + m_B v_{B2}$$
$$v_{B2} - v_{A2} = e\,(v_{A1} - v_{B1}).$$

An important strategy is to assume both v_{A2} and v_{B2} are moving in the positive direction in both equations. In this way we will know the direction of a particle is negative when the results are negative. In Example 11.2 we demonstrate that the direction of the results can change depending on the value of CoR.

Example 11.2
Ball A ($m_A = 2$ kg) moves ($v_{A1} = 6$ m/s) toward ball B ($m_B = 6$ kg) moving ($v_{B1} = 4$ m/s) in the opposite direction as shown in Figure 11.8. Determine the velocities of the balls and the

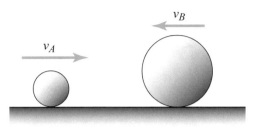

Figure 11.8: **Example** 11.2.

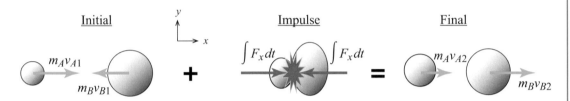

Figure 11.9: **Example** 11.2 Impulse-Momentum diagram.

efficiencies if the CoR varies from $e = 0$ to $e = 1$ in increments of 0.25. Graph the results of efficiency vs. CoR.

Note: We assume the positive direction for both resulting velocities (Figure 11.9).

$$\sum m \vec{v}_1 = \sum m \vec{v}_2$$

x-dir:

$$m_A v_{A1} - m_B v_{B1} = m_A v_{A2} + m_B v_{B2}$$

$$(2)(6) - (6)(4) = (2) v_{A2} + (6) v_{B2}$$

$$(-12) = (2) v_{A2} + (6) v_{B2} \; \textcircled{1}$$

CoR:

$$v_{B2} - v_{A2} = e (v_{A1} - v_{B1})$$

$$v_{B2} - v_{A2} = e ((6) - (-4))$$

$$v_{B2} - v_{A2} = e (10) \; \textcircled{2} \, .$$

Combining equations $\textcircled{1}$ and $\textcircled{2}$, we can find the relations:

$$v_{A2} = \frac{(-12) - e(60)}{(8)} \quad \text{and} \quad v_{B2} = \frac{(12) - e(20)}{(4)}.$$

If we plug in the varying CoRs we get the results in Table 11.1.

Table 11.1: Velocities after collision in Example 11.2

e	v_{A2}	v_{B2}
0	-1.500	-1.500
0.25	-3.375	-0.875
0.5	-5.250	-0.250
0.75	-7.125	0.375
1	-9.000	1.000

Table 11.2: Velocities after collision in Example 11.2

e	KE_2	η
0	9.000	10.7%
0.25	13.69	16.3%
0.5	27.75	33.0%
0.75	51.19	60.9%
1	84.00	100%

We note that ball A always reverses direction, going to the left (negative x-direction). Ball B keeps moving to the left until the CoR increases enough so it bounces backward to the right. We calculate that at $e = 0.6$ ball B will simply stop since $0 = \frac{(12)-e(20)}{(4)}$.

The efficiency can be found from the ratio of kinetic energy:

$$\eta = \frac{WYW}{WYPF} = \frac{KE_2}{KE_1}.$$

Energy Before:

$$KE_1 = \frac{1}{2}m_A v_{A1}^2 + \frac{1}{2}m_B v_{B1}^2 = \frac{1}{2}(2)(6)^2 + \frac{1}{2}(6)(-4)^2 = 84.00 \text{ Nm}.$$

Energy After:

$$KE_2 = \frac{1}{2}m_A v_{A2}^2 + \frac{1}{2}m_B v_{B2}^2 = \frac{1}{2}(2) v_{A2}^2 + \frac{1}{2}(6) v_{B2}^2.$$

We present the results in Table 11.2 and Figure 11.10.

We can see there is a relationship between CoR and Efficiency, but they are not the same thing.

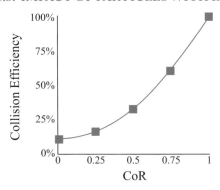

Figure 11.10: Example 11.2 efficiency vs. CoR.

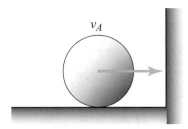

Figure 11.11: Ball vs. wall.

11.3 IMPACT OF PARTICLES WITH IMMOVABLE BARRIERS

We often want to analyze a particle striking a large, immovable barrier, such as a wall (Figure 11.11) or the ground, but applying the Conservation of Momentum presents a problem. Common sense tells us the momentum of the barrier is zero, but if we try to calculate it we use zero velocity (it's not moving) and a mass that approaches infinite. Zero multiplied by infinity is either zero or indeterminant, therefore we can't use the barrier's momentum in the analysis. To demonstrate this draw the Impulse-Momentum diagram to the scenario where a ball strikes a wall (Figure 11.12).

If we attempt to apply the Conservation of Momentum and set the wall's momentum to zero, we get the non-sensical result $v_{A1} = v_{A2}$. This approach should show the ball reverses direction. Therefore, we conclude this won't work.

$$\sum m\vec{\mathbf{v}}_1 = \sum m\vec{\mathbf{v}}_2$$

$$m_A v_{A1} + m_B v_{B1} = m_A v_{A2} + m_B v_{B2}$$

$$m_A v_{A1} + \infty\,(0) = m_A v_{A2} + \infty\,(0).$$

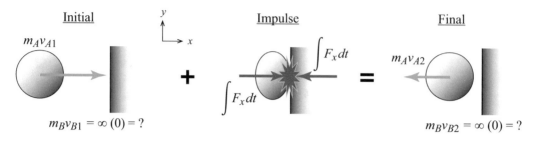

Figure 11.12: Ball vs. wall Impulse-Momentum diagram.

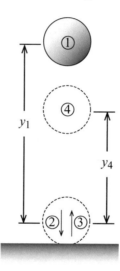

Figure 11.13: Example 11.3.

So what do we do? We'll apply the CoR instead, noting that the velocity of the wall is zero:

$$v_{B2} - v_{A2} = e\,(v_{A1} - v_{B1})$$

$$(0) - v_{A2} = e\,(v_{A1} - (0))$$

$$v_{A2} = -e v_{A1}.$$

The result here says the velocity will reverse and be less which makes sense. Example 11.3 demonstrates a scenario we could test ourselves.

Example 11.3
A ball is dropped from a height of 6 ft (Figure 11.13). How high will it bounce if the CoR is $e = 0.8$? What is the efficiency of this bounce?

We first find the velocity of the ball as it strikes the ground using kinetics:

$$v_2{}^2 = v_1{}^2 - 2g\,(y_2 - y_1)$$
$$v_2{}^2 = (0)^2 - 2\,(32.2)\,((0) - (6))$$
$$v_2 = 19.66 \text{ ft/s}.$$

Using the CoR we find the rebound velocity:

$$v_{G3} - v_{B3} = e\,(v_{B2} - v_{G2})$$
$$(0) - v_{B3} = e\,(v_{B2} - (0))$$
$$v_{B3} = -ev_{B2}$$
$$v_{B3} = -\,(0.8)\,(19.66) = 15.73 \text{ ft/s}.$$

The rebound height is found from kinematics:

$$v_4{}^2 = v_3{}^2 - 2g\,(y_4 - y_3)$$
$$(0)^2 = (15.73)^2 - 2\,(32.2)\,(y_4 - (0))$$
$$y_4 = 3.840 \text{ ft}$$
$$\boxed{y_4 = 3.84 \text{ ft}}.$$

Energy Before:

$$PE_1 = mgy_1 = mg\,(6) \qquad \text{(This is "what you paid for")}.$$

Energy After:

$$PE_4 = mgy_4 = mg\,(3.840) \qquad \text{(This is "what you want")}.$$

$$\eta = \frac{WYW}{WYPF} = \frac{PE_4}{PE_1} = \frac{mg\,(3.840)}{mg\,(6)} = 64.00\%$$
$$\boxed{\eta = 64.0\%}.$$

We note that the mass of the ball is not provided and wasn't needed.

We see that the rebound loses energy and the ball doesn't return to the original height. We could use this in the reverse manner to determine both the CoR and efficiency of various balls by dropping them from a known height and measuring the return height. Recording a slow motion video with a tape measure in the frame will help measuring the balls' return height.

Golf clubs are regulated by the USGA to have a CoR not greater than 0.83 (for most clubs, drivers use a different standard).

http://www.golfclub-technology.com/coefficient-of-restitution.html

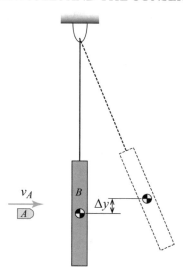

Figure 11.14: Example 11.4.

11.4   BALLISTIC PENDULUM

A "ballistic pendulum" is a method to measure the speed of a projectile. The projectile is propelled toward a block hung from a rope. After the collision, the height the block rises to is measured and that change in potential energy is used to estimate the kinetic energy (and therefore speed) of the projectile. This is often done with bullets which embed into the block (CoR $e = 0$). Example 11.4 demonstrates a ballistic pendulum where the projectile doesn't embed.

Example 11.4

A 0.25-oz bullet travels toward a 10 lb steel plate supported by a cable and starting at rest. The CoR is $e = 0.75$. The plate rises to a maximum height of $\Delta y = 1$ inch. Determine the speed of the bullet before and after it hits the plate (Figure 11.14).

We'll call the plate in motion (just after the bullet strikes it) state "2" and the final position of the plate (when it pauses at the top) state "3".

From Conservation of Energy, using state 2 as the datum:

$$KE_2 + PE_2 = KE_3 + PE_3$$

$$\frac{1}{2}m_B v_{B2}^2 + (0) = (0) + m_B g \Delta y.$$

Notably, the mass cancels:

$$\frac{1}{2}v_{B2}^2 + (0) = (0) + (32.2)\left(\frac{1}{12}\right)$$

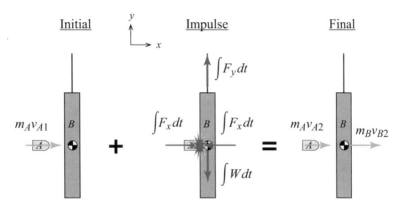

Figure 11.15: Impulse-Momentum diagram of Example 11.4.

$$v_{B2} = 2.317 \text{ ft/s}.$$

Figure 11.15 shows the Impulse-Momentum diagram for this example. In the x-direction we can use Conservation of Linear Momentum if we take both the bullet and the plate since the impulse is internal to this system:

$$\sum m\vec{\mathbf{v}}_1 = \sum m\vec{\mathbf{v}}_2$$

x-dir:

$$m_A v_{A1} + m_B v_{B1} = m_A v_{A2} + m_B v_{B2}$$

$$\left(\frac{0.25/16}{32.2}\right) v_{A1} + \left(\frac{10}{32.2}\right) (0) = \left(\frac{0.25/16}{32.2}\right) v_{A2} + \left(\frac{10}{32.2}\right) (2.317)$$

$$(4.852E - 4) v_{A1} = (4.852E - 4) v_{A2} + (0.7196)$$

$$v_{A2} = v_{A1} - (1{,}483) \ ①.$$

We have one equation and two unknowns. We apply the coefficient of restitution as the other equation.

CoR:

$$v_{B2} - v_{A2} = e (v_{A1} - v_{B1})$$

$$(2.317) - v_{A2} = (0.75)(v_{A1} - (0))$$

$$v_{A2} = (2.317) - (0.75) v_{A1} \ ②.$$

Equating equations ① and ②:

$$v_{A1} - (1{,}483) = (2.317) - (0.75) v_{A1}$$

$$v_{A1} = 848.7 \text{ ft/s}.$$

The question doesn't ask for this, but typically high speeds are presented in mph since we are quite familiar with magnitudes in that unit:

$$v_{A1} = \frac{(848.7 \text{ ft/s}) (3600 \text{ s/hr})}{(5280 \text{ ft/mi})} = 578.6 \text{ mph}$$

$$\boxed{v_{A1} = 579 \text{ mph}} \; .$$

The rebound (ricochet) speed of the bullet is:

$$v_{A2} = (2.317) - (0.75)(848.7) = -634.2 \text{ ft/s} = 432.4 \text{ mph} \leftarrow \; .$$

We also should take note that we made certain assumptions about the behavior of the plate. First, we treated it like a particle which means its rotation isn't considered. This kept the plate in line with the cable which we can imagine wouldn't be true unless it was a very stiff cable attached to the plate in a way that wouldn't allow it to pivot. In a real ballistic pendulum, the object might be supported by multiple cables so rotation isn't a factor or it might be a rigid connection and therefore require rigid body motion calculations, as is covered in Class 24 (vol. 2).

Book 1 - Class 12

https://www.youtube.com/watch?v=waHV1ghuktI

C L A S S   12

Oblique Impact of Particles

B.L.U.F. (Bottom Line Up Front)

- Oblique Impact uses normal (aligned by the center of particles) and tangential coordinates.

- Tangential velocity remains constant for each particle.

- Conservation of momentum is applied in the normal direction as if it were a direct impact problem.

- Coefficient of Restitution in the normal direction is: $e = \frac{(v_{B2})_n - (v_{A2})_n}{(v_{A1})_n - (v_{B1})_n}$.

12.1   OBLIQUE IMPACT

In the previous Class we covered impact where particles move along a straight line. We call this "direct impact" (sometimes also referred to as "direct central impact") as shown in Figure 12.1a. Now we ask what is "indirect impact"? That is, what happens when the particles' velocities aren't along the same line as in Figures 12.1b and c. Note the line drawn through the particle centers also includes the point where they touch. We call this the "line of impact."

To organize the analysis of an oblique impact we create a coordinate system where the line of impact is the normal direction and the tangential direction is perpendicular to it. As shown in Figure 12.2, we break the velocity vectors into components in this coordinate system.

Note the normal direction velocity components align, so in this direction we have a direct central impact. An Impulse-Momentum diagram of this collision is shown in Figure 12.3.

We recall from Class 10 that Impulse Momentum uses vector equations, so each direction can therefore be treated separately.

We also recall from Class 10 that Impulse Momentum can be applied to a system of particles or individual particles. Figure 12.4 shows an Impulse-Momentum diagram for each particle.

For the individual particles we see we cannot perform the analysis in the normal direction because we don't know the value of the impulse. In the tangential direction, however, we see there is no impulse, so we can use the conservation of momentum. The results show us:

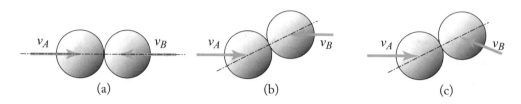

Figure 12.1: Particle impacts: (a) direct, (b) oblique directions aligned, and (c) oblique directions not aligned.

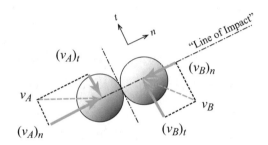

Figure 12.2: Oblique impact coordinates and velocity components of (c) in Figure 12.1.

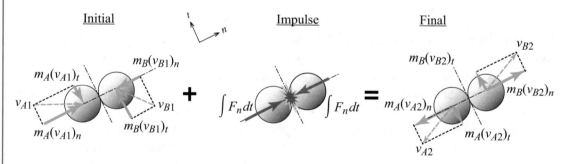

Figure 12.3: Oblique impact Impulse-Momentum diagram of multiple particles.

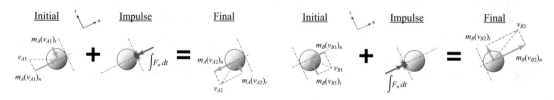

Figure 12.4: Oblique impact Impulse-Momentum diagram of individual particles.

$$m_A (v_{A1})_t = m_A (v_{A2})_t$$
$$m_B (v_{B1})_t = m_B (v_{B2})_t.$$

Our conclusion is that $(v_{A1})_t = (v_{A2})_t$ and $(v_{B1})_t = (v_{B2})_t$. So the velocity in the tangential direction is unchanged.

There *could* be an impulse in the tangential direction *if* the balls had friction. But most often we assume the balls are very smooth and therefore the impact frictionless.

In the normal direction we see from Figure 12.3 that the impulse cancels out, and we can use the Conservation of Linear Momentum, just like the direct central impact from Class 11. We need to revise the CoR to make it specific to the normal direction velocities:

$$\boxed{e = \left| \frac{(v_{B2})_n - (v_{A2})_n}{(v_{A1})_n - (v_{B1})_n} \right|}.$$

A general approach to a two particle oblique impact analysis is as follows.

1. Draw the Impulse-Momentum diagram of the two balls.

2. Create a normal and tangential coordinate system.

3. Break the initial velocities into normal and tangential components.

4. Apply the Conservation of Momentum to *each particle* individually in the tangential direction, resulting in $(v_{A1})_t = (v_{A2})_t$ and $(v_{B1})_t = (v_{B2})_t$.

5. Apply the Conservation of Momentum to the *system* in the normal direction to get an equation unknown $(v_{A2})_n$ and $(v_{B2})_n$. Assume positive final direction for both.

6. Apply the CoR relationship in the normal direction (again, assume positive final direction) to get a second equation with these unknowns and solve.

7. Combine the components from four and six components to find the final speeds and directions.

8. Don't forget to account for the rotation of the coordinate system when answering with the actual angles.

Billiards (pool), like Newtdog is playing with Wormy in Figure 12.5, is a common application of oblique impact, especially when "cutting" a ball to make it go toward a pocket or hit another ball. This is usually done by picking a point on the second ball and aiming the cute ball at that spot. We'll see how that works in Example 12.1. We'll also demonstrate other common applications of oblique impact:

Figure 12.5: Newtdog plays billiards with Wormy (repeat of Figure 11.1) (© E. Diehl).

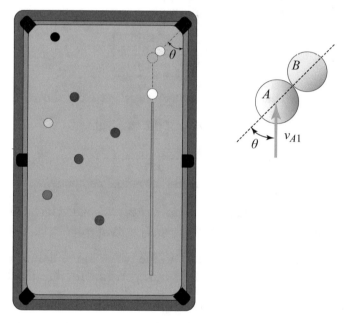

Figure 12.6: Example 12.1.

https://www.real-world-physics-problems.com/physics-of-billiards.html
https://hypertextbook.com/facts/2004/OluwoleOwoseni.shtml

Example 12.1

The cue ball ($m_A = 0.17$ kg) is struck toward the 1-ball ($m_B = 0.16$ kg) at 10 m/s on a felt pool table with negligible friction. The angle between the balls is $\theta = 45°$, as shown in Figure 12.6. The coefficient of restitution between the two balls is $e = 0.9$. Determine the *velocity* of each ball after impact (magnitude and angle with respect to horizontal of the page) and the efficiency of the collision.

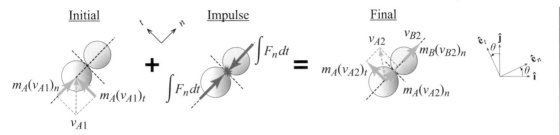

Figure 12.7: Example 12.1 Impulse-Momentum diagram.

For this first example we'll label the steps.

Steps 1 and 2: Impulse-Momentum diagram (Figure 12.7) with normal and tangential coordinates.

Step 3: Break the initial velocity into normal and tangential components.

$$(v_{A1})_t = (10) \sin\left(45°\right) = 7.701 \text{ m/s} \searcharrow$$
$$(v_{A1})_n = (10) \cos\left(45°\right) = 7.701 \text{ m/s} \nearrow .$$

Apply Conservation of Momentum

$$\sum m\vec{\mathbf{v}}_1 = \sum m\vec{\mathbf{v}}_2$$

Step 4: In the t-direction.

$$m_A \left(v_{A1}\right)_t = m_A \left(v_{A2}\right)_t \qquad \left(v_{A1}\right)_t = \left(v_{A2}\right)_t = 5.000 \text{ m/s}$$
$$m_B \left(v_{B1}\right)_t = m_B \left(v_{B2}\right)_t \qquad \left(v_{B1}\right)_t = \left(v_{B2}\right)_t = 0.$$

Step 5: In the n-direction.

$$m_A \left(v_{A1}\right)_n + m_B \left(v_{B1}\right)_n = m_A \left(v_{A2}\right)_n + m_B \left(v_{B2}\right)_n$$

Assume positive direction of A for final

$$(0.17)(7.071) + (0.16)(0) = (0.17)\left(v_{A2}\right)_n + (0.16)\left(v_{B2}\right)_n$$
$$(1.202) = (0.17)\left(v_{A2}\right)_n + (0.16)\left(v_{B2}\right)_n \,\, ① .$$

Step 6: CoR

$$e = \frac{\left(v_{B2}\right)_n - \left(v_{A2}\right)_n}{\left(v_{A1}\right)_n - \left(v_{B1}\right)_n} = \frac{\left(v_{B2}\right)_n - \left(v_{A2}\right)_n}{(7.071) - (0)} = 0.9$$

Table 12.1: Transformation matrix

	$\hat{\imath}$	$\hat{\jmath}$
$\hat{e}_t$	$-\sin\theta$	$\cos\theta$
$\hat{e}_n$	$\cos\theta$	$\sin\theta$

$$(v_{B2})_n - (v_{A2})_n = (6.364)\ \text{\textcircled{2}}$$

$$(1.202) = (0.17)(v_{A2})_n + (0.16)(6.364) + (0.16)(v_{A2})_n.$$

Solve ① and ②:

$$(v_{A2})_n = 0.5568\ \text{m/s} \qquad (v_{B2})_n = 6.921\ \text{m/s}.$$

Step 7: Combine the components.
Velocity magnitudes:

$$v_{A2} = \sqrt{(v_{A2})_t^2 + (v_{A2})_n^2} = \sqrt{(7.701)^2 + (0.5568)^2} = 7.721\ \text{m/s}$$

$$v_{B2} = \sqrt{(v_{B2})_t^2 + (v_{B2})_n^2} = \sqrt{(0)^2 + (6.921)^2} = 6.921\ \text{m/s}.$$

Step 8: Transform the results into horizontally aligned coordinates (Table 12.1).

$$\vec{v}_{A2} = (7.701)\,\hat{e}_t + (0.5568)\,\hat{e}_n$$

$$\vec{v}_{B2} = (0)\,\hat{e}_t + (6.921)\,\hat{e}_n.$$

$$\vec{v}_{A2} = \left[-(7.701)\sin(45°) + (0.5568)\cos(45°)\right]\hat{\imath}$$
$$+ \left[(7.701)\cos(45°) + (0.5568)\sin(45°)\right]\hat{\jmath}$$

$$\vec{v}_{A2} = (-5.052)\,\hat{\imath} + (5.839)\,\hat{\jmath}\ \text{m/s}$$

$$\theta_A = \tan^{-1}\frac{(5.839)}{(5.052)} = 49.13°\ \text{CW from horizontal}$$

$$\vec{v}_{B2} = \left[-(0)\sin(45°) + (6.921)\cos(45°)\right]\hat{\imath}$$
$$+ \left[(0)\cos(45°) + (6.921)\sin(45°)\right]\hat{\jmath}$$

$$\vec{v}_{B2} = (4.894)\,\hat{\imath} + (4.894)\,\hat{\jmath}\ \text{m/s}$$

$$\theta_B = \tan^{-1} \frac{(4.894)}{(4.894)} = 45° \text{ (this is an obvious result)}$$

$$\boxed{\vec{v}_{A2} = 7.72 \text{ m/s} \searrow 49.1°} \qquad \boxed{\vec{v}_{B2} = 6.92 \text{ m/s} \nearrow 45°} .$$

To find the efficiency we apply Work-Energy:

$$KE_1 + PE_1 + U_{1 \rightarrow 2} = KE_2 + PE_2$$

where

$$KE_1 = \tfrac{1}{2} m v_{A1}^2 + \tfrac{1}{2} m v_{B1}^2 = \tfrac{1}{2} (0.17) (10)^2 + \tfrac{1}{2} (0.16) (0)^2 = 8.500 \text{ J}$$

$$PE_1 = 0$$

$$U_{1 \rightarrow 2} = -Losses$$

$$KE_2 = \tfrac{1}{2} m v_{A2}^2 + \tfrac{1}{2} m v_{B2}^2 = \tfrac{1}{2} (0.17) (7.093)^2 + \tfrac{1}{2} (0.16) (6.921)^2 = 8.108 \text{ J}$$

$$PE_2 = 0$$

$$(8.500) + (0) - Losses = (8.108) + (0) .$$

Efficiency: $\eta = \frac{Initial\ energy - Losses}{Initial\ energy} = \frac{KE_2}{KE_1} = \frac{(8.108)}{(8.500)} = 95.39\%$

$$\boxed{\eta = 95.4\%} .$$

Example 12.2

The cue ball from Example 12.1 continues at 7.721 m/s $\searrow$ 49.13° and banks off the rail as shown in Figure 12.8. The coefficient of restitution between the ball and rail is $e = 0.6$. Determine the velocity (speed and direction) of the cue ball after the collision and the efficiency of the collision.

Steps 1 and 2: We draw the Impulse-Momentum diagram (Figure 12.9) even though we recall that we can't use the Conservation of Linear Momentum in the normal direction. Drawing the diagram helps us think through the collision velocity components. In this case, the normal direction aligns with the vertical of the page and the tangential the horizontal. In this case we know the normal component of the ball's velocity will reverse direction so we draw it this way.

Step 3: Break the initial velocity into normal and tangential components

$$(v_{A1})_n = (7.721) \sin (49.13°) = 5.839 \text{ m/s} \uparrow$$
$$(v_{A1})_t = (7.721) \cos (49.13°) = 5.052 \text{ m/s} \longleftarrow .$$

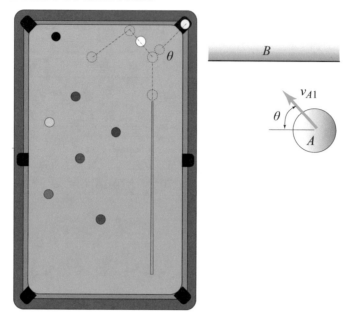

Figure 12.8: Example 12.2.

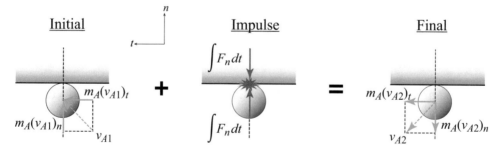

Figure 12.9: Example 12.2 Impulse-Momentum diagram.

Apply Conservation of Momentum

$$\sum m\vec{\mathbf{v}}_1 = \sum m\vec{\mathbf{v}}_2.$$

Step 4: We can apply the Conservation of Linear Momentum in the tangential direction to the ball alone and conclude:

$$m_A\,(v_{A1})_t = m_A\,(v_{A2})_t \qquad (v_{A1})_t = (v_{A2})_t = 5.052 \text{ m/s} \longleftarrow .$$

Step 5: We can't apply the Conservation of Linear Momentum in the normal direction because the rail is unmovable (has zero velocity and essentially infinite mass, leaving it indeterminant).

Step 6: In the normal direction we can use the CoR:

$$e = \frac{(v_{B2})_n - (v_{A2})_n}{(v_{A1})_n - (v_{B1})_n} = \frac{(0) + (v_{A2})_n}{(5.839) - (0)} = 0.6$$

$$(v_{A2})_n = (0.6)(5.839) = 3.503 \text{ m/s} \downarrow .$$

Step 7: Combine the components:

$$v_{A2} = \sqrt{(v_{A2})_t^2 + (v_{A2})_n^2} = \sqrt{(5.052)^2 + (3.503)^2} = 6.148 \text{ m/s}$$

$$\theta_B = \tan^{-1} \frac{(3.503)}{(5.052)} = 34.74° .$$

Step 8: The components are already aligned with the page coordinates.

$$\boxed{\vec{\mathbf{v}}_{A2} = 6.15 \text{ m/s} \nearrow 34.7°} .$$

To find the efficiency we apply Work-Energy:

$$KE_1 + PE_1 + U_{1 \to 2} = KE_2 + PE_2$$

where

$$KE_1 = \tfrac{1}{2}mv_{A1}^2 = \tfrac{1}{2}(0.17)(7.721)^2 = 5.067 \text{ J}$$

$$PE_1 = 0$$

$$U_{1 \to 2} = -Losses$$

$$KE_2 = \tfrac{1}{2}mv_{A2}^2 = \tfrac{1}{2}(0.17)(6.148)^2 = 3.213 \text{ J}$$

$$PE_2 = 0$$

$$(5.067) + (0) - Losses = (3.213) + (0)$$

Efficiency: $\eta = \frac{Initial\ energy - Losses}{Initial\ energy} = \frac{KE_2}{KE_1} = \frac{(3.213)}{(5.067)} = 63.40\%$

$$\boxed{\eta = 63.4\%} .$$

The low CoR of the ball and rail indicates that energy is lost/absorbed during the collision, and this efficiency confirms it.

Figure 12.10: Example 12.3 (repeat of Figure 10.3) (© E. Diehl).

Example 12.3

We revisit Example 10.4 where a baseball was hit out of the ballpark. Now we want to investigate the bat and ball interaction like Newtdog hitting Wormy's apple in Figure 12.10. Recall the ball in Example 10.1 was pitched at 95 mph (139.3 ft/s horizontally), and its weight was 5 oz ($9.705 \cdot 10^{-3}$ slugs). We concluded that the ball had to travel 105.7 ft/s horizontally and 76.13 ft/s vertically (for an angle of 35.76° from horizontal) to land out of the park. We treat the baseball bat as a particle (B) with a weight of 33 oz ($6.405 \cdot 10^{-2}$ slugs) and swung horizontally, striking the ball with an oblique angle. The CoR of bat on ball is estimated to be $e = 0.6$. Determine the speed of the bat to achieve these results.

We won't label the steps for this example, but we are still using them. Figure 12.11 is the Impulse-Momentum diagram for the baseball bat and ball.

We write out the given velocities in Cartesian coordinates:

$$\vec{v}_{A1} = (-139.3\,)\,\hat{\imath}\ \text{ft/s}$$

The initial velocities:

$$\vec{v}_{B1} = v_{B1}\hat{\imath}$$

$$\vec{v}_{A2} = (105.7)\,\hat{\imath} + (76.13)\hat{\jmath}\ \text{m/s}$$

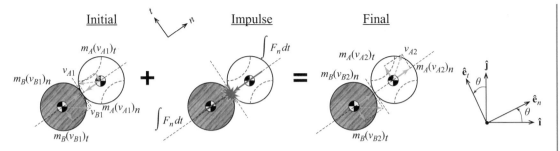

Figure 12.11: Impulse-Momentum diagram of Example 12.3.

Table 12.2: Transformation matrix of Example 12.3

	$\hat{\mathbf{i}}$	$\hat{\mathbf{j}}$
$\hat{\mathbf{e}}_t$	$-\sin\theta$	$\cos\theta$
$\hat{\mathbf{e}}_n$	$\cos\theta$	$\sin\theta$

The known velocities need to be transformed into normal and tangential coordinates using the unit vectors shown in Figure 12.3 to create the Transformation matrix in Table 12.2. We don't know the angle to transform it yet, so we'll keep this as an unknown parameter, theta:

$$\vec{\mathbf{v}}_{A1} = -(-139.3)\sin\theta\,\hat{\mathbf{e}}_t + (-139.3)\cos\theta\,\hat{\mathbf{e}}_n$$

$$\vec{\mathbf{v}}_{A1} = (139.3)\sin\theta\,\hat{\mathbf{e}}_t + (-139.3)\cos\theta\,\hat{\mathbf{e}}_n \text{ ft/s}$$

$$\vec{\mathbf{v}}_{B1} = -v_{B1}\sin\theta\,\hat{\mathbf{e}}_t + v_{B1}\cos\theta\,\hat{\mathbf{e}}_n$$

$$\vec{\mathbf{v}}_{A2} = [-(105.7)\sin\theta + (76.13)\cos\theta]\,\hat{\mathbf{e}}_t + [(105.7)\cos\theta + (76.13)\sin\theta]\,\hat{\mathbf{e}}_n.$$

We know the velocity in the tangential direction of each particle remains the same before and after the impact. We can set the tangential coordinates of particle A equal to find the necessary line-of-action angle:

$$(v_{A1})_t = (v_{A2})_t$$

$$(v_{A1})_t = (139.3)\sin\theta = (v_{A2})_t = -(105.7)\sin\theta + (76.13)\cos\theta$$

$$(139.3)\sin\theta = -(105.7)\sin\theta + (76.13)\cos\theta$$

$$\frac{\sin\theta}{\cos\theta} = \tan\theta = \frac{(76.13)}{(139.3) + (105.7)}$$

$$\theta = 17.26°$$

Apply Conservation of Momentum

$$\sum m \vec{\mathbf{v}}_1 = \sum m \vec{\mathbf{v}}_2.$$

In the n-direction:

$$(v_{B1})_t = (v_{B2})_t = -v_{B1} \sin \theta$$

$$m_A (v_{A1})_n + m_B (v_{B1})_n = m_A (v_{A2})_n + m_B (v_{B2})_n$$

$$\left(9.705 \cdot 10^{-3}\right) (-139.3) \cos \left(17.26°\right) + \left(6.405 \cdot 10^{-2}\right) v_{B1} \cos \left(17.26°\right)$$
$$= \left(9.705 \cdot 10^{-3}\right) \left[(105.7) \cos \left(17.26°\right) + (76.13) \sin \left(17.26°\right)\right] + \left(6.405 \cdot 10^{-2}\right) (v_{B2})_n$$

$$(-1.29103) + \left(6.117 \cdot 10^{-2}\right) v_{B1} = (1.19884) + \left(6.405 \cdot 10^{-2}\right) (v_{B2})_n$$

$$(v_{B2})_n = (-38.87) + (0.9550) \, v_{B1} \; ①.$$

CoR:

$$e = \frac{(v_{B2})_n - (v_{A2})_n}{(v_{A1})_n - (v_{B1})_n} = \frac{(v_{B2})_n - (105.7) \cos \left(17.26°\right) + (76.13) \sin \left(17.26°\right)}{(-139.3) \cos \left(17.26°\right) - v_{B1} \cos \left(17.26°\right)} = 0.6$$

$$(v_{B2})_n - (105.7) \cos \left(17.26°\right) + (76.13) \sin \left(17.26°\right)$$
$$= (0.6) (-139.3) \cos \left(17.26°\right) - (0.6) \, v_{B1} \cos \left(17.26°\right)$$

$$(v_{B2})_n = (-1.4645) - (0.5730) \, v_{B1} \; ②.$$

Solve ① and ②:

$$(-38.87) + (0.9550) \, v_{B1} = (-1.4645) - (0.5730) \, v_{B1}$$

$$v_{B1} = 24.48 \text{ ft/s}$$

$$\boxed{\vec{\mathbf{v}}_{B1} = 24.5 \text{ ft/s} \; \rightarrow}.$$

This is a demonstration of applying oblique impact, but the real world is a bit more complicated. For instance, we're assuming that the spin of the ball doesn't matter but it does. Check out some of these informative links on the physics of baseball to get a better appreciation for the complications one might want to consider to do a thorough analysis:

http://baseball.physics.illinois.edu/oblique.html
http://www.physics.usyd.edu.au/~cross/baseball.html

Example 12.4
Ball A ($W_A = 5$ lb) is released from rest at the top ($y_1 = 8$ ft) of a slope ($\theta = 35°$) that ends at $y_2 = 6$ ft. It strikes block B ($W_B = 20$ lb) which is moving upward at its maximum speed at

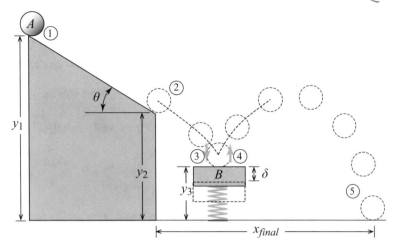

Figure 12.12: Example 12.4.

$y_3 = 2$ ft after having had its spring ($k = 200$ lb/ft) compressed δ and released. The block and ball impact with perfect timing to send the ball through the air to land at $x_{final} = 40$ ft as shown in Figure 12.12. The coefficient of restitution between the ball and block is $e = 0.8$. Determine the amount the block needs to be compressed (δ) to achieve this distance.

This problem purposely has multiple concepts, so we have to think through and then recognize the principles we need. We'll first need to know the velocities from Work-Energy and Projectile Motion so we can analyze the oblique impact of the ball and block. We numbered the locations/events on Figure 12.12 to keep track of them.

Work-Energy of the ball from $1 \to 2$:

$$KE_1 + PE_1 + U_{1\to 2} = KE_2 + PE_2$$

where

$KE_1 = 0$ ball starts from rest

$PE_1 = W_A y_1 = (5)(8) = 40$ ft · lb

$U_{1\to 2} = 0$ (assume the slope is frictionless)

$KE_2 = \frac{1}{2}m_A v_{A2}^2 = \frac{1}{2}\left(\frac{5}{32.2}\right)v_{A2}^2 = \left(7.764 \cdot 10^{-2}\right)v_{A2}^2$

$PE_2 = W_A y_2 = (5)(6) = 30$ ft · lb

$$(0) + (40) + (0) = \left(7.764 \cdot 10^{-2}\right)v_{A2}^2 + (30)$$

$$v_{A2} = 11.35 \text{ ft/s} \searrow 30°.$$

We might be tempted to find the velocity at point 3 directly using Work-Energy, but remember this would only give us the magnitude and we need to know the components:

$$(v_{A2})_x = v_{A2} \cos \theta = (11.35) \cos \left(30°\right) = 9.829 \text{ ft/s} \rightarrow$$

$$(v_{A2})_y = v_{A2} \sin \theta = (11.35) \sin \left(30°\right) = 5.675 \text{ ft/s} \downarrow .$$

From Projectile Motion:

$$(v_{A3})_y^2 = (v_{A2})_y^2 - 2g \left(y_3 - y_2\right)$$

$$(v_{A3})_y^2 = (5.675)^2 - 2 \left(32.2\right) \left((2) - (6)\right)$$

$$(v_{A3})_y = 17.02 \text{ ft/s} \downarrow .$$

The time from $2 \rightarrow 3$:

$$(v_{A3})_y = (v_{A2})_y - g t_{2\rightarrow3}$$

$$(-17.02) = (-5.675) - (32.2) t_{2\rightarrow3}$$

$$t_{2\rightarrow3} = 0.3524 \text{ s.}$$

The horizontal position change from $2 \rightarrow 3$:

$$\Delta x_{2\rightarrow3} = (v_{A2})_x \, t_{2\rightarrow3} = (9.829) (0.3524) = 3.464 \text{ ft.}$$

The time from $4 \rightarrow 5$ (leaving block to landing zone):

$$\Delta x_{4\rightarrow5} = (v_{A2})_x \, t_{4\rightarrow5}$$

$$(20 - 3.464) = (9.829) t_{4\rightarrow5}$$

$$t_{4\rightarrow5} = 1.682 \text{ s.}$$

The required vertical velocity at point 4 is:

$$y_5 = y_4 + (v_{A4})_y \, t_{4\rightarrow5} - \frac{1}{2} g t_{4\rightarrow5}^2$$

$$(0) = (2) + (v_{A4})_y (1.682) - \frac{1}{2} (32.2) (1.682)^2$$

$$(v_{A4})_y = 25.90 \text{ ft/s} \uparrow .$$

We draw the Impulse-Momentum diagram of the ball and block for the impact analysis in Figure 12.13.

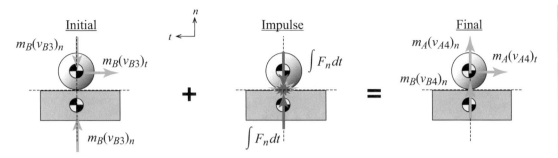

Figure 12.13: Impulse-Momentum diagram of Example 12.4.

Apply Conservation of Momentum

$$\sum m\vec{\mathbf{v}}_3 = \sum m\vec{\mathbf{v}}_4$$

$$m_A\,(v_{A3})_n + m_B\,(v_{B3})_n = m_A\,(v_{A4})_n + m_B\,(v_{B4})_n$$

$$\left(\frac{5}{32.2}\right)(-17.02) + \left(\frac{20}{32.2}\right)(v_{B3})_n = \left(\frac{5}{32.2}\right)(25.90) + \left(\frac{20}{32.2}\right)(v_{B4})_n$$

Assume both move in positive direction at 4

$$(v_{B4})_n = (v_{B3})_n - (10.73)\ \ \textcircled{1}.$$

CoR:

$$e = \frac{(v_{B4})_n - (v_{A4})_n}{(v_{A3})_n - (v_{B3})_n} = \frac{(v_{B4})_n - (25.90)}{(-17.02) - (v_{B3})_n} = 0.8$$

$$(v_{B4})_n = (12.28) - (0.8)\,(v_{B3})_n\ \ \textcircled{2}$$

$$(v_{B3})_n - (10.73) = (12.28) - (0.8)\,(v_{B3})_n$$

$$(v_{B3})_n = 12.79\ \text{m/s}\ \uparrow.$$

We will re-use the "position 2" designation for when the block is compressed and apply Work-Energy:

Work-Energy of the block from $2 \rightarrow 3$:

$$KE_2 + PE_2 + U_{2\rightarrow3} = KE_3 + PE_3$$

where:

$KE_2 = 0$ block starts from rest

$PE_2 = \frac{1}{2}k\delta^2 = \frac{1}{2}\,(12 \cdot 200)\,\delta^2 = (1{,}200)\,\delta^2$

$U_{2 \to 3} = 0$ no losses

$KE_3 = \frac{1}{2} m_B v_{B3}^2 = \frac{1}{2} \left(\frac{20}{32,2} \right) (12.79)^2 = 50.77 \text{ ft} \cdot \text{lb}$

$PE_3 = W_B \delta = (20) \delta$

$$(0) + (1,200) \delta^2 + (0) = (50.77) + (20) \delta$$

$$(1,200) \delta^2 - (20) \delta - (50.77) = 0$$

$$\delta = \frac{-(-20) \pm \sqrt{(-20)^2 - 4 (1,200) (-50.77)}}{2 (1,200)} = 0.2142, \ -0.1975 \text{ ft}$$

(negative result is meaningless)

Answer: $\boxed{\delta = 0.214 \text{ ft} = 2.57 \text{ in}}$.

12.2 PARTICLE DYNAMICS CONCLUSIONS

In conclusion: Dynamics is hard. The only way to get better at solving Dynamics problems is to practice solving Dynamics problems. Reading the solution to examples won't help as much as working through the examples on your own. Start with easy examples, cover up the answers, and solve them all the way through until you arrive at an answer and only then check your results against the solution. Keep doing this and push yourself and you will not only get better at solving Dynamics problems, you'll get better at solving other engineering problems. Following this routine is the secret to success in engineering courses.

This course companion is continued in *Part 2: Rigid Bodies, Kinematics, and Kinetics*.

In Figure 12.14, Newtdog and Wormy say "So long," "Rock on," and "Go chill on a beach somewhere" after finals week.

Figure 12.14: **Newtdog and Wormy out.**

A P P E N D I X   A

Particle Dynamics Sample Exam Problems

This appendix provides sample problems for Exams 1 (Particle Kinematics) and 2 (Particle Kinetics) corresponding to the course sechdule in Table 0.1. These are included so students can practice for the exams with problems of the approximate level of difficulty they might expect. A typical 75-min exam might include three to four of these problems, so students are encouraged to time themselves when taking them and attempt to stay under approximately 20 min each.

A.1   PARTICLE KINEMATICS SAMPLE EXAM PROBLEMS

These problems are covered in Classes 1–5.

A.1.1   PROJECTILE MOTION 1

According to a recent magazine article, one of the big hitters on a major golf tour (initials D.J.) typically has a driving launch angle and ball speed of 12.2° and 186.5 mph, respectively. On a level fairway, how many yards will his ball carry before landing if air resistance is neglected? (1 yard = 3 ft). See Figure A.1.

`https://www.golfdigest.com/story/these-dustin-johnson-launch-monitor-`
`numbers-are-absolutely-ridiculous`

A.1.2   PROJECTILE MOTION 2

The most famous golfer in the world has been measured to average $v_0 = 180$ mph ball speed. His new sponsor has developed a new driver that can achieve a $\theta = 15°$ launch angle. If the elevated green on the golf course he's playing is $h = 20$ ft above the tee box, how many yards away should he be to sink the drive on the fly, thus scoring the elusive par 4 albatross (also known as a double eagle). (1 yard = 3 ft). See Figure A.2.

A.1.3   PARTICLE KINEMATICS 1

An airplane is at the bottom of a curved dive in the vertical plane. Its speed is 400 ft/s which is increasing at a rate of 10 ft/s². The radius of curvature of the path is 2,500 ft. The plane is being tracked by radar using polar coordinates from the location shown. See Figure A.3.

At this instant the radar is measuring:

Figure A.1: Example practice problem A.1.1.

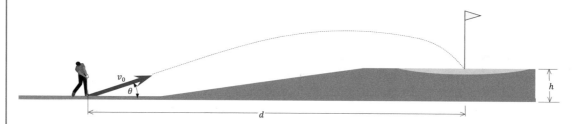

Figure A.2: Example practice problem A.1.2.

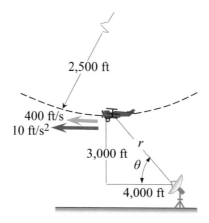

Figure A.3: Example practice problem A.1.3.

(a) $\ddot{r}$.

(b) $\ddot{\theta}$.

A.1.4   PARTICLE KINEMATICS 2

Starting from rest, a sprinter runs outward with a constant acceleration of 1.5 ft/s² in the radial direction from the center of a platform. The platform is rotating at a constant angular velocity $\dot{\theta} = 0.2$ rad/s. See Figure A.4. When he has traveled 12 ft what is:

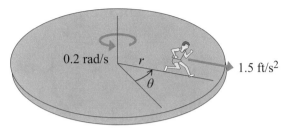

Figure A.4: Example practice problem A.1.4.

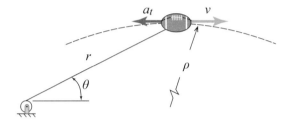

Figure A.5: Example practice problem A.1.5.

(a) The *speed* of the sprinter.

(b) The *magnitude of the acceleration* of the sprinter.

A.1.5   PARTICLE KINEMATICS 3

One end of a fishing line is attached to a thrown football which is at its peak in the instant shown. The football's speed is 75 ft/s, slowing at 2 ft/s^2 due to air resistance in high winds. The radius of curvature of the football path is $\rho = 175$ ft. The fishing line is $r = 60$ ft at $\theta = 20°$ from horizontal. See Figure A.5.

(a) What is the rate (ft/s) of the fishing line spooling out?

(b) What is the angular velocity (rad/s) of the fishing line?

(c) What is the magnitude of the total acceleration (ft/s^2)?

A.1.6   PARTICLE KINEMATICS 4

A fuel cell-powered car prototype is found to accelerate along a test track (Figure A.6) by the relation $a = 6 - 0.002 \cdot s$ m/s^2, where s is the distance the car travels along the track starting from Point A, where it begins at rest. The distance from Point A to Point B is $x = 100$ m and the radius of the track ends is $r = 50$ m. Find the acceleration when the car reaches Point C, providing the magnitude and angle with respect to horizontal, indicating quadrant.

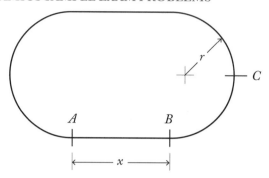

Figure A.6: Race track for problem A.1.6.

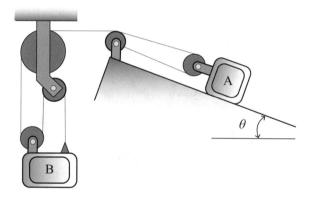

Figure A.7: Blocks and pulleys for problem A.1.7.

A.1.7   PARTICLE KINEMATICS 5

The searchlight on a stationary hovering helicopter located $d = 2000$ ft from a highway is track-ing a race car traveling at $v = 60$ mph and increasing speed at 15 ft/s². At the instant shown in Figure A.7 the distance of the searchlight to the race car is $r = 3000$ ft. Determine the angular acceleration required for the searchlight to rotate while following the car.

A.1.8   PARTICLE KINEMATICS 6

Block A is positioned on a $\theta = 20°$ sloped surface and connected via inextensible rope and pulleys to block B, as shown in Figure A.8. The system starts at rest and after the block has moved up the slope 2.5 m, block B is traveling at $\vec{v}_B = 3$ m/s ↓. Determine the relative acceleration of block B with respect to block A, providing the magnitude and angle with respect to horizontal, indicating quadrant.

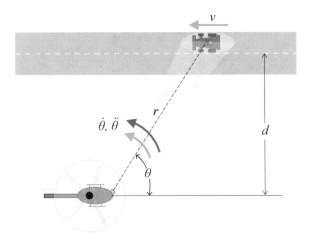

Figure A.8: Helicopter tracking race car in problem A.1.8.

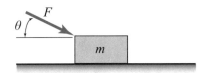

Figure A.9: Example practice problem A.2.1.

A.2   PARTICLE KINETICS SAMPLE EXAM PROBLEMS

These problems are covered in Classes 6–12.

A.2.1   PARTICLE N2L 1

The box shown has a mass of 20 kg. A 400 N force is applied at an angle $\theta = 35°$. The coefficients of friction are $\mu_s = \mu_k = 0.30$. See Figure A.9. You are *required* to draw and label the FBD and IBD for this problem.

Determine:

(a) The *acceleration* of the box.

(b) The *velocity* of the box after it has traveled 12 m, starting from rest.

A.2.2   PARTICLE N2L 2

Block A ($m_A = 10$ kg) is connected to block B ($m_B = 50$ kg) on a $\theta = 20°$ slope by an inextensible cable via massless and frictionless pulleys. All surfaces are frictionless. A force ($F = 100$ N)

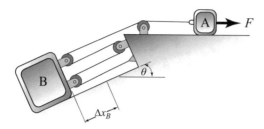

Figure A.10: Example practice problem A.2.2.

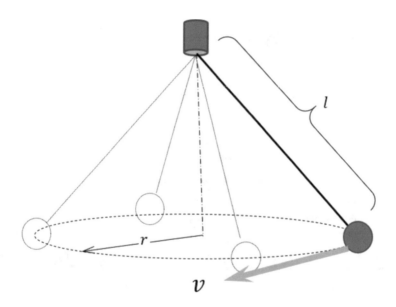

Figure A.11: Example practice problem A.2.3.

is applied to block A and movement begins. See Figure A.10. Two properly labeled FBD and IBD sets are required as part of the answer to this problem. Determine the velocity (m/s) of block B after it moves up the slope $\Delta x_B = 2$ m.

A.2.3   PARTICLE N2L 3

A device spins a 1 kg ball attached to a 400-mm-long string in a 200-mm-radius circle as shown. See Figure A.11. What constant ball speed will result in this configuration? Both an FBD and IBD (properly labeled) are required as part of the answer to this problem.

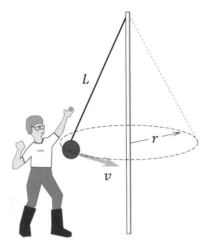

Figure A.12: Example practice problem A.2.4.

A.2.4 PARTICLE N2L 4

A regulation tetherball is 0.5 kg with a 0.6-m-long rope. The tetherball rope has a proof load of 20 N. See Figure A.12. How fast (m/s) does the ball need to travel to break the rope? A properly labeled FBD and IBD set are required as part of the answer to this problem.

A.2.5 PARTICLE N2L 5

The 10-lb block A is at rest when the 20-lb block B is released and both begin to move. The coefficient of kinetic friction is $\mu_k = 0.2$. See Figure A.13. Two FBD and IBD sets (properly labeled) are required as part of the answer to this problem. After release:

(a) What is the acceleration of block A?

(b) What is the acceleration of block B?

(c) What is the tension in the rope?

A.2.6 WORK-ENERGY 1

Block A ($m_A = 25$ kg) is connected to block B ($m_B = 45$ kg) by an inextensible cable and massless, frictionless pulley. Both are at rest and begin to move when a $\vec{F} = 25$ N $\nearrow$ 30° force is applied to Block A. A spring ($k = 10$ N/m) is connected to block A and is at its free length (no compression or extension) in the initial position. The coefficient of friction between block A and the surface is $\mu = 0.25$. See Figure A.14.

Determine:

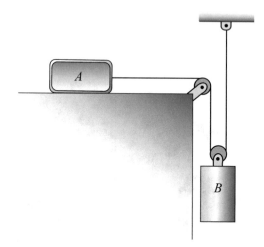

Figure A.13: Example practice problem A.2.5.

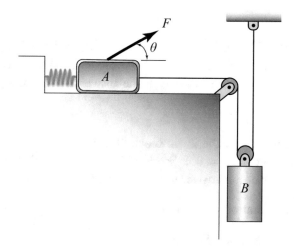

Figure A.14: Example practice problem A.2.6.

(a) The *velocity* of block B when it has moved 1.5 m down using **Work-Energy**.

(b) Will the tension in the cable: be constant, increase, or decrease during this motion? Explain why.

A.2.7   WORK-ENERGY 2

Block A ($m_A = 20$ kg) on a slope ($\theta = 60°$) is connected to block B ($m_B = 40$ kg) by an inextensible cable and massless, frictionless pulley. Both are released from rest and begin to move.

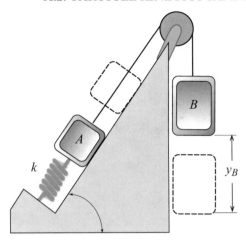

Figure A.15: Example practice problem A.2.7.

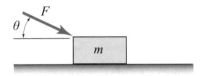

Figure A.16: Example practice problem A.2.8.

A spring ($k = 10$ N/m) connected to block A is at its free length (no compression or extension) in the position shown. The coefficient of friction between block A and the sloped surface is $\mu = 0.2$. See Figure A.15. Using Work-Energy, determine the velocity of block B when it has moved 3 m down.

A.2.8   IMPULSE-MOMENTUM 1

The box from exam 1 is back. Its mass is $m = 10$ kg, and the applied force is $F = 100$ N to the right at $\theta = 25°$ from horizontal as shown. See Figure A.16. The coefficients of friction are $\mu_s = \mu_k = 0.25$. If the box is already moving at $v_1 = 50$ m/s before the force is applied, what is its speed after the force has been applied for $\Delta t = 10$ s? (Use linear impulse and momentum to solve this problem.)

A.2.9   IMPULSE-MOMENTUM 2

A spaceship ($m = 500$ kg) is traveling at 1,000 m/s in a straight line along the x-axis. The pilot will apply the right and rear thrusters to change velocity to $\vec{v} = 2000$ m/s $\nearrow$ 30°, where the angle is measured CCW from the x-axis. The front and rear thrusters produce constant 15 kN

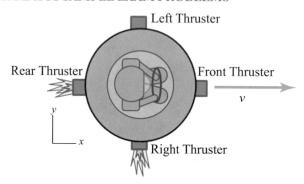

Figure A.17: Example practice problem A.2.9.

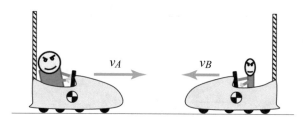

Figure A.18: Example practice problem A.2.10.

while the side thrusters produce constant 10 kN. See Figure A.17. Determine the *times* each thruster must be applied, Δt_x and Δt_y.

A.2.10   PARTICLE IMPACT 1

A big kid and his little brother are playing "chicken" with bumper cars at the amusement park. The older brother and his car weigh 800 lb driving toward his brother at $v_A = 10$ ft/s. The little brother, whose car and he weigh 750 lb together, drives toward his brother at $v_B = 40$ ft/s. The coefficient of restitution between bumper cars is $e = 0.5$. See Figure A.18.
   Determine:

(a) The *velocity* of each brother after the collision.

(b) The efficiency of the collision.

A.2.11   PARTICLE IMPACT 2

A big kid is bullying his little brother while driving bumper cars at the amusement park. He and his car have a mass $m_A = 400$ kg which he drives toward his brother at $v_A = 10$ m/s. The little brother, whose car and he have a mass $m_B = 350$ kg, futilely tries to escape at $v_B = 5$ m/s. The

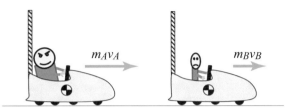

Figure A.19: Example practice problem A.2.11.

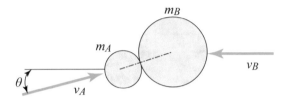

Figure A.20: Example practice problem A.2.12.

coefficient of restitution between bumper cars is $e = 0.5$. See Figure A.19. Find the *velocity* of each brother after the collision.

A.2.12 OBLIQUE PARTICLE IMPACT

A small ball ($m_A = 20$ kg) approaches a larger ball ($m_B = 40$ kg) at an angle ($\theta = 30°$) as shown. The small ball's speed is $v_A = 10$ m/s, and the larger ball's speed is $v_B = 5$ m/s. The coefficient of restitution is $e = 0.75$. See Figure A.20. Determine the *speed* of each ball after the collision.

A.3   ANSWERS TO SAMPLE EXAM PROBLEMS

Table A.1 presents the answers to the sample exam problems. It is almost certain that most students will be disappointed that fully worked out solutions are not provided. I encourage you to do the difficult work of solving the problems and persevering until you find where you went astray (that is, if you don't get the same results the first time you attempt). This extra effort will pay off in your exam preparation as finding your own mistakes helps to avoid making them again.

Table A.1: Answers to sample exam problems

A.1.1 - Projectile Motion 1	320 yards
A.1.2 - Projectile Motion 2	334 yards
A.1.3 - Particle Kinematics 1	$\ddot{r} = 57.9$ ft/s^2, $\ddot{r} = 0.0152$ rad/s^2
A.1.4 - Particle Kinematics 2	$|\vec{v}| = 6.46$ ft/s, $|\vec{a}| = 2.61$ ft/s^2
A.1.5 - Particle Kinematics 3	$\dot{r} = 70.5$ ft/s, $\dot{\theta} = 0.428$ rad/s, $|\vec{a}| = 32.2$ rad/s^2
A.1.6 - Particle Kinematics 6	$\vec{a} = 42.0$ rad/s^2 $\measuredangle$ 7.73°
A.1.7 - Particle Kinematics 7	$\ddot{\theta} = 4.19E - 3$ rad/ s^2 ↺
A.1.8 - Particle Kinematics 8	$\vec{a}_{B/A} = 5.58$ m/s^2 ↘ 47.0°
A.2.1 - Particle N2L 1	$\vec{a} = 10.0$ m/s^2 → , $\vec{v} = 15.5$ m/s →
A.2.2 - Particle N2L 2	$\vec{v}_B = 2.10$ m/s $\measuredangle$ 20°
A.2.3 - Particle N2L 3	$v = 1.06$ m/s
A.2.4 - Particle N2L 4	$v = 4.75$ m/s
A.2.5 - Particle N2L 5	$\vec{a}_A = 17.2$ ft/s^2 →, $\vec{a}_B = 8.59$ ft/s^2 ↓, $F_T = 7.33$ lb
A.2.6 - Work-Energy 1	$\vec{v}_{B2} = 2.65$ m/s ↓
A.2.7 - Work-Energy 2	$\vec{v}_B = 4.33$ m/s ↓
A.2.8 - Impulse-Momentum 1	$\vec{v}_2 = 106$ m/s →
A.2.9 - Impulse-Momentum 2	$\Delta t_x = 24.4$ s, $\Delta t_y = 50.0$ s
A.2.10 - Particle Impact 1	$v_{A2} = -26.3$ ft/s, $v_{B2} = -1.29$ ft/s, $\eta = 43.3\%$
A.2.11 - Particle Impact 2	$\vec{v}_A' = 6.50$ m/s →, $\vec{v}_B' = 9.00$ m/s →
A.2.12 - Oblique Particle Impact	$|v_A'| = 6.72$ m/s, $|v_B'| = 4.74$ m/s

Author's Biography

EDWARD DIEHL

Dr. Edward Diehl obtained his doctoral degree in Mechanical Engineering from the University of Connecticut in December 2016. He is currently an Assistant Professor at the University of Hartford in the Mechanical Engineering Department. Prior to joining UHartford, he was a lecturer (2009–2017) at the United States Coast Guard Academy in both the Mechanical Engineering section and Naval Architecture and Marine Engineering section. He worked as a Principal Engineer (2006–2009, 1996–2000, and 1992–1995) for Seaworthy Systems, Inc., self-employed (2000–2006), and an analyst (1995–1996) for General Dynamics/Electric Boat. He is a registered Professional Engineer in Connecticut. He obtained a Master of Science in Mechanical Engineering from Rensselaer at Hartford in 1996. He is a proud graduate of the United States Merchant Marine Academy at Kings Point, class of 1992, with a Bachelor of Science degree in Marine Engineering Systems. His research interests include solid mechanics pedagogy, gear vibration and fault modeling, and mechanism design.

Printed in the United States
by Baker & Taylor Publisher Services